# BIOLOGICAL WAR

# ENCYCLOPAEDIA OF BIOTERRORISM

## Vol. I

## BIOLOGICAL WAR

*By*

**Dr. S.K. Prasad**

*School of Studies of Zoology & Biotechnology*

*Vikram University*

*Ujjain*

**DISCOVERY PUBLISHING HOUSE PVT. LTD.**

**NEW DELHI-110 002**

First Published-2008

ISBN 978-81-8356-302-4 (Set)



*Published by:*

**DISCOVERY PUBLISHING HOUSE PVT. LTD.**
4831/24, Ansari Road, Prahlad Street,
Darya Ganj, New Delhi-110002 (India)
Phone: 23279245 • Fax: 91-11-23253475
*E-mail:* dphbooks@rediffmail.com
dphtemp@indiatimes.com

*Printed at:*
Sachin Printers, Delhi

# Preface

The present title *"Encyclopaedia of Bioterrorism"* aims to bring historical context to present concerns about biological weapons, biological agents, chemical weapons and the potential for bioterrorism. The lack of use of biological weapons in war advocates for using biology to create a new class of weapons initially envisioned delivery systems for pathogenic aerosols that mimicked those for chemical weapons, which were mainly bombs that generated aerosols intended to kill or disable troops in a local area. This vision was quickly replaced by the concept of creating huge clouds of germs that would drift with the wind and infect people over areas of thousands of square miles. The scientists and civil and military leaders who believed in the future of biological weapons saw their potential for fulfilling the goals of total war, for the mass killing or debilitation of enemy civilians. It is often asked why biological weapons are different from any other means of destruction. The answer is that they are the only ones devised expressly to kill defenseless humans and animal and plant life, with little real battlefield potential in modern war.

When biological weapons were developed for possible retaliation against an enemy thought to be similarly armed, they fit this model of restraint. Nevertheless, germ weapons were developed and laws and political circumstances offered no guarantees against their use. Historians may seek neat explanations, but the element of uncertainty was always there. Political and military authorities differed unpredictably when it came to calculating the consequences of using biological weapons. This book also describes the new subject of bioterrorism and traces at least the beginnings of the present era, in which domestic preparedness and homeland security are major policy issue. One of the major homeland security directives is to use technology derived from medical research to protect civilians against bioterrorism. The use of biology to defined civilians against any and all biological agents is a daunting project that imposes new security restrictions more familiar to physicists in the defense establishment than to biologists.

The study of biological weapons combines knowledge from disparate fields: biology, medicine, military history, politics, law, and ethics. This book intends to give the reader a basic literacy in this complex area. The entire subject of biological weapons is characterized by an unusual degree of misinformation and even disinformation. Almost every fact in this book has several page backstory, with greater nuance and depth than a brief overview can provide. As scholars continue their work, more information and analysis will likely turn today's accepted wisdom on its head. This progress will be a healthy sign for the field, whose subject matter has often been exploited for the frightening and sometimes entertaining effect it has on the imagination.

There can be no claim to originality except in the manner of treatment and much of the information has been obtained from the books and scientific journals available in the different libraries.

The author expresses his thanks to his friends and colleagues whose continue inspirations have initiated him to bring out this book.

The author expresses his gratitude to Mr. Wasan and staff of M/s Discovery Publishing House for their whole hearted co-operation in the publication of this book.

**Author**

# Contents

**1. Biological Warfare** **1—18**

Spies and Sabotage, Man of Secrets, Biological Warfare and Science Advisors, Sir Frederick Banting, "Fog of War", Biological Weapons Funding from Industrial Giants, Biological Weapons Experiments and Testing, First Field Tests.

**2. Biological Threat and Atomic Bomb** **19—39**

Biological Warfare, the Atomic Bomb, and the Soviet Threat, Biological Weapons Startup, Technical Restraints, Korean War Accusations, Toward Improved Testing, Operation Harness, St Jo Program, Project Whitecoat, Options for First Strike, Expansion in the Kennedy-Johnson Years.

**3. Threat Created by Bioweapons** **40—52**

Prevention Bioterrorist Attacks, Threat Created by the Bioweapons, Protection of Food from Bioterrorists, Are We Prepared for a Bioterrorist Attack? Bioterrorist Attack, Key Areas in Public Health, Training, Communications and Coordination, Special Focus, From a Dark Winter to a Bright Future, Response to a Bioterrorist Attack, Challenges.

**4. Airborne Diseases** **53—71**

Diseases of the Upper Respiratory Tract, Streptococcal Diseases, Diphtheria, Pertussis, Meningococcal Meningitis, Haemophilus Meningitis, Diseases of the Lower Respiratory Tract, Tuberculosis, Pneumococcal Pneumonia, Primary Atypical Pneumonia (PAP), Klebsiella Pneumonia, Serratia Pneumonia, Legionnaires' Disease (Legionellosis), Q Fever, Psittacosis and Chlamydial Pneumonia.

5. **Foodborne and Waterborne Diseases** **72—90**

Foodborne and Waterborne Intoxications, Botulism, Staphylococcal Food Poisoning, Clostridial Food Poisoning, Foodborne and Waterborne Infections, Typhoid Fever, Salmonellosis, Shigellosis, Cholera, E. coli Diarrheas, Peptic Ulcer Disease, Campylobacteriosis, Brucellosis, Other Foodborne and Waterborne Diseases.

6. **Soilborne Diseases** **91—107**

Soilborne Bacterial Diseases, Anthrax, Tetanus, Gas Gangrene, Leptospirosis, Listeriosis, Melioidosis, Arthropodborne Bacterial Diseases, Bubonic Plague, Tularemia, Lyme Disease, Relapsing Fever, Rickettsial Arthropodborne Diseases, Rocky Mountain Spotted Fever, Epidemic Typhus, Endemic Typhus.

7. **Physical Control** **108—120**

Physical Control with Heat, Direct Flame, Hot-Air Oven, Boiling Water, Autoclave, Fractional Sterilization, Pasteurization, Hot Oil, Physical Control by Other Methods, Filtration, Ultraviolet Light, Other Types of Radiation, Ultrasonic Vibrations, Preservation Methods.

8. **Chemical Control** **121—135**

General Principles of Chemical Control, Terminology of Disinfection, Selection of Antiseptics and Disinfectants, Evaluation of Antiseptics and Disinfectants, Important Chemical Agents, Halogens, Phenol and Phenolic Compounds, Heavy Metals, Alcohols, Other Chemical Agents, Formaldehyde, Ethylene Oxide, Glutaraldehyde, Hydrogen Peroxidase, Soaps and Detergents, Dyes, Acids.

9. **Antibiotics** **136—152**

Chemotherapeutic Agents, A Short History of Chemotherapy, Sulfanilamide and Other Sulfonamides, Other Chemotherapeutic Agents, Antibiotics, Discovery of Antibiotics, Penicillin, Semisynthetic Penicillins, Cephalosporins, Aminoglycosides, Chloramphenicol, Other Antibiotics, Tetracyclines, Miscellaneous Antibiotics, Antifungal Antibiotics, Antibiotic Assays and Resistance, Antibiotic Sensitivity Assays, Antibiotic Resistance and Abuse.

10. **Biological Challenge** **153—173**

Life Sciences Today, Dual Use Dilemma, Brief History of Modern Biological Warfare, U.S. Policy and the Creation of the Biological and Toxin Weapons Convention, New Threat, Recent Examples of "Contentious Research" in the Life Sciences, Mousepox Virus: A Case Study in Preconsideration, Total Synthesis of the Poliovirus

Genome and Recovery of Infectious Virus, Comparison of the Immune Response to a Virulence Gene from Vaccinia and Smallpox, Response of the Life Sciences Community to Previous Challenges, Committee Charge and Process, Committee Process, Biological Warfare in History.

**11. Regulatory Environment** **174—198**

The U.S. Regulatory Environment, Oversight of Genetic Engineering Research, Framework for Implementation of the NIH Guidelines for rDNA Research, Types of Experiments That Require IBC, RAC, and NIH Director Review, Physical and Biological Containment Strategies for NIH-funded rDNA Research Activities, Compliance with and Enforcement of NIH Guidelines, Regulation of Microbial Agents (Listed Agents and Toxins), Possession of Select Agents, Importation and Interstate Shipment of Etiologic Agents, Oversight of Foreign Nationals, Additional Security Checks on Visa Applications, Tracking Systems, Professional Education and Responsibilities of Life Scientists, International Situation, Laboratory Safety, Research Oversight, Conclusions.

**12. Sensitive Information** **199—214**

Past as Prologue? Nuclear Weapons Complex, Cryptography, Lessons from the Comparisons, Secret and Sensitive Information, Secret Information, Sensitive Information, Defining "Sensitive" Information, Dual Use Information, Publication of Sensitive Information in the Life Sciences.

**13. Recommendations** **215—256**

Key Assumptions, Many More Experiments will Explore the Virulence Factors of Bacteria and Viruses, Scientific Evaluation of the Risks is Essential, Only an International Set of Standards will Help to Minimize the Misuse of Biotechnology, Recommendations, Educating the Scientific Community, Review of Plans for Experiments, Review at the Publication Stage, Creation of a National Science Advisory Board for Biodefense, Additional Elements for Protection Against Misuse, A Role for the Life Sciences in Efforts to Prevent Bioterrorism and Biowarfare, Harmonized International Oversight, Conclusion, National Policy on NBC Warfare Requires Updating, DoD NBC Defense RDA Program is Still Recovering, CB Agents and Domestic Terrorism, There is a Future for the Chemical Corps.

**Index** **257—266**

# 1

# Biological Warfare

In 1940, the British government embarked on a biological warfare program that began serendipitously and yet established an enduring organizational model of laboratory research and field tests of munitions. The UK initiative pioneered the way for the American effort, which then greatly surpassed it in size but not in ingenuity. The British program presents an example of "*inadvertent escalation*" under circumstances of war, with high officials initially unaware that research on a new type of weapon had been authorized.[1] British dread of Germany's destructive power motivated the decision to prepare retaliatory biological warfare capacity. Earlier suspicion of Germany's capacity and intent had already aroused British concerns about the threat of biological weapons.

## Spies and Sabotage

German attempts to infect the Allies' pack animals with glanders and anthrax in World War I were minor exploits compared to the implications of the 1934 revelations of British journalist Henry Wickham Steed. According to secret documents Steed said he had received, German agents had investigated the potential impact of aerial attacks of chemical and biological weapons on London and Paris.

The German use of airplanes and airships (especially the zeppelin) in World War I had caused dread among the British population. These German bombings killed 1,400 British civilians, a small number compared to British battlefield losses but a terrifying demonstration that the English Channel no longer served as a reliable protective barrier against traditional enemies on the continent. The distinction between combatant and noncombatant, already blurred by industrial war, could dissolve in air war, where the inaccurate bombing of military targets and the targeting of cities put civilians in double jeopardy. Steed's revelations, suggesting Germany's possible combination of airplanes and deadly aerosols, raised concern throughout Europe.

Wickham Steed began his article by translating an alleged July1932 memo on strategic bombing from the commander of a secret German department, *Air Gas Defense* (*Luft-Gas-verteidigung*): "As you were informed some time ago, the gigantic French fortifications on our Western frontier make attack by infantry seem quite, and artillery attack almost, hopeless. Consequently there remains only the most intensive development and extension of the air weapon, in order that air warfare may be waged effectively and ruthlessly against important military and industrial centres and, above all, also against the civilian population of large cities."

Other German documents quoted by Steed outlined plans for spraying a usually harmless red bacteria, *Bacillus prodigiosus*, at entrances to the Paris Metro and the London Underground. This use of a simulant the pathogen now commonly known as *Serratia marcescens*, made sense as a testing device. As presented, the idea was to track how air currents might spread toxic gas or pathogens through the entire subway system over, for example, a six-hour period. The premise was that bacteria or chemicals from a low-flying aircraft would be sucked into the ventilation systems of underground transportation. The subway systems in Berlin and Hamburg had already been tested, the documents claimed, in the event of an enemy attack by air. The documents also presented details about specific August 18, 1933, tests outside eight Paris Metro stops.

Not all experts were convinced of the authenticity of Steed's information. The documents contained internal inconsistencies and scientific notations that were not in German or even recognizably scientific. One critic pointed out that, after six hours, the stiff winds in Paris on August 18 would have blown the dispersed bacteria as far as the Belgian frontier. Steed defended the authenticity of the papers he had received, on the basis of his sources, which he kept anonymous.

The clumsiness of the alleged German tests took nothing from their alarming intent. In public, British and French governments played down the significance of the Wickham Steed documents. Behind closed doors, British intelligence added Steed's information to another report received around the same time. Its spies had reported that an ongoing Berlin course for advanced military specialists in gas weapons included material on bacteriological war fare. According to these sources, the teachers of the course insisted that Germany would only use biological agents in retaliation, thus adhering to the Geneva Protocol. This intelligence was deemed inconclusive, but the seeds of suspicion about a German biological weapons threat had been planted.

In this period, faulty intelligence on both sides led to misapprehensions of the enemy's chemical and biological warfare (CBW) strengths and intentions. At times intelligence was misjudged or ignored and government suppositions left much to fate. Given a reliable account from a German prisoner of war that the Third Reich had developed a nerve gas, tabun, the British government dismissed the information. For their part, the Germans mistakenly believed the British chemical warfare program had also developed nerve gas, a belief that apparently deterred their own use of tabun, which they had manufactured by the ton. After the war, Reichsmarshall Hermann Goering told his captors that the Nazi command had resisted using nerve gas during the D-Day invasion for fear that an Allied retaliation in kind would kill the precious pack animals on which the German forces, low on gasoline for trucks and jeeps, were heavily reliant.

In the buildup to World War II, the British and their Allies continued to receive intermittent intelligence reports about German biological weapons production, all of them suggestive and none, as it turned out, accurate. Adolf Hitler's personal aversion to biological weapons was a major restraint. German microbiologists, who speculated that the British might spray anti- crop diseases or airdrop vials of lethal bacteria, were limited to small aerosol experiments. In addition to Hitler's objections, during the Third Reich the great legacy of German biology degenerated to eugenics and racist theories that led to the death-camp experiments and murders of Jews and Gypsies. Until the end of 1944, when it was certain that the Germans had no biological warfare capacity, the British justified their program's retaliatory goals with "*looking-glass*" presumptions that whatever their biological weapons pro gram achieved, the Germans must also have accomplished.

## Man of Secrets

The British might have confined themselves to purely defensive measures against biological warfare—public health reinforcement and, for the military, vaccines, drugs, and face masks—had it not been for Maurice Hankey, longtime secretary to the British Cabinet and the most influential civil servant in the interwar years.

Hankey's 1974 biography Man of Secrets makes no reference to his activities on behalf of the program, which continued after the war, and his collected personal papers do not reveal his crucial influence. After years of historians representing him as antagonistic to the biological weapon program, Hankey was eventually acknowledged as its founding father by UK biological weapons scientists. Newly declassified documents, reported on by British historian Brian Balmer, also document the centrality of Hankey's hidden role.

Throughout his career, Hankey was a behind-the-scenes influence on British defense policy. During World War I, he was Secretary to the War Cabinet. In 1918, he became Secretary to the British Cabinet, a position he invented and then held for twenty years, while simultaneously acting as Secretary of the Committee of Imperial Defence. Hankey organized Cabinet and Defence agendas and consulted closely with Prime Minister Nevile Chamberlain in the years leading up to World War II. He served on numerous select committees and created others that he often chaired or otherwise controlled. Officially retired in 1938, Hankey returned to government to aid the mobilization of the United Kingdom for the coming conflict with Germany. In 1939, he became Minister without Portfolio in the new War Cabinet, whose organization he had planned in detail, and he supported the successful proposal for a Ministry of Home Security to prepare defenses against German air raids. He also worked closely with leaders of another new bureaucracy, the Ministry of Economic Warfare, which spawned Special Operations Executive, a clandestine agency responsible for sabotage operations and later a client of the biological weapons program.

## Biological Warfare and Science Advisors

Before the Wickham Steed documents were made public, they were given to Hankey, who brought them to the attention of three eminent scientists, John Ledingham, director of the Lister Institute, Professor William Topley of the London School of Hygiene and Tropical Medicine, and Captain Stewart Douglas, deputy director of the National Institute for Medical Research. All three were ambivalent about the authenticity of the documents, they thought the tests, if they actually occurred, related to chemical and not biological agent dispersal. Furthermore, the three scientists were skeptical about bio logical weapons in general. Unlike Auguste Trillat, whose work they may not have known they believed that the shock from explosive devices would kill most bacteria and that sprays from airplanes would result in aerosols too diffuse to cause serious, large-scale outbreaks.

In the aftermath of the Wickham Steed controversy, Hankey met privately with physiologist Sir Edward Mellanby, secretary of the Medical Research Council, to ask whether the Council might address the biological warfare threat. Mellanby emphatically rejected the idea. He saw the Council's mission as the promotion of medical research for healing, not for military ends, and certainly not to prepare weapons, even for retaliation.

Rather than retreat, Hankey cultivated Mellanby as an advisor and encouraged his project to improve the nutritional status of army recruits. In 1936, Hankey became chair of the Subcommittee on Bacteriological Warfare, with in the Imperial Defense Ministry. He recruited Mellanby as a member, and also Ledingham and Topley and other top medical scientists. The task of the committee was to report on "the practicality of *biological weapons*" and recommend "countermeasures which should be taken to deal with such an eventuality." Mellanby and his scientific colleagues began meeting with representatives from the armed forces and from the *Chemical Defence Research Establishment*, most of whose budget went to the chemical weapons research program at Porton Down, west of London on the relatively remote Wiltshire plain.

As late as 1938, the science advisors Hankey had assembled doubted the practicality of germ weapons and therefore the German biological weapons threat. Instead, the committee's main concern was with the potential casual ties from German air raids on the British and, secondarily, with the possible diseases that might result from the injuries, disruptions, and hardships of war. The most influential committee members had firsthand experience with large-scale epidemics. They had served overseas—in Mesopotamia, Serbia, India, and China—under conditions of war and social disruption that had generated serious outbreaks. They were also of the generation that had experienced the 1918 influenza epidemic, which, as it circled the globe killing millions, demonstrated the opportunistic nature of epidemics. The committee concentrated on what it perceived as future wartime civilian vulnerability in the United Kingdom: the public health needs brought on by wounds front aerial bombings, threats to water supplies, and possible mass epidemics among civilians debilitated by wartime food and housing deprivations.

Writing about the effects of total war on Great Britain, historian Arthur Marwick describes 1938 as a period when "politicians were obsessed with the problem of civilian casualties." This obsession became the driving force for the creation of the Emergency Hospital Service, which required cooperation between voluntary and local hospitals regionally and nationally. By 1941, 80 percent of all British hospitals were integrated into the scheme and they opened their doors for free care to a wide variety of patients in need, not just bomb casualties and veterans. The Emergency Hospital Service became an integral part of the postwar framework for nationalized health care.

Attuned to public health, Hankey's committee prepared a long report proposing a national network of emergency bacteriological laboratories to aid in better diagnosis and treatment of civilian casualties. The report was passed to the Medical Research Council, headed by Mellanby, which established a special subcommittee to evaluate it and appointed Topley, one of its authors, as committee chair. In July 1938, the Committee of Imperial Defence approved the proposal for a national laboratory, which Mellanby named the Emergency Public Health Laboratory Service. This network of laboratories endured after the war as the Public Health Laboratory Service, an important part of disease management in the United Kingdom.

Until this juncture, the work of Hankey's committee had been highly secret. Now the members appealed to Hankey for permission to advertise their public health objectives openly. The Committee of Imperial Defence responded by allowing only limited contacts with nongovernmental organizations involved with the committee's other project, the stockpiling of pharmaceuticals.

After September, 1939, when Britain and Germany were officially at war, Hankey pressed his committee of advisors to focus on the technical potentials of biological weapons. While its members admitted that sabotage could frighten and demoralize the public and that botulinum toxin could be especially dangerous, they still rejected the notion that biological weapons could cause a serious epidemic. They were unimpressed by the increasing but fragmentary intelligence reports of German biological weapons technology: bombs, glass vials, and sprays. When asked to review the practicality of different methods, they stuck to their skepticism about the threat of aerosols and put their faith in routine, efficient public health responses. They discounted the dangers of *anthrax*, *tularemia*, and other noncontagious diseases. In a November 1939 report, Mellanby and the committee empathically rejected the airplane as "a very clumsy and inefficient bacteriological instrument" in capable of "depositing an infective agent in the right place at the right time."

Nonetheless, Hankey persisted in questioning his science advising corn mittee about scientific uncertainties regarding disease agents. Had the reactions of bacteria to explosives been tested? How might anthrax spores travel through the air? Finally, Mellanby asked Hankey to get permission from Prime Minister Chamberlain for a series of aerosol experiments at the Nations Institute for Medical Research. Chamberlain agreed, after Hankey represented the experiments as purely defensive, in no way intended for weapons development. These tests, never fully reported, could have been either defensive or offensive in their purpose. Mellanby's request proved to be a first step toward more comprehensive exploration of the potential of anthrax and other potential biological weapons agents.

## Sir Frederick Banting

Meanwhile, in Canada, Frederick Banting, head of his own research center at the University of Toronto, had become convinced that Hitler and the German army, as well as the Italians and the Japanese, would wage biological warfare and that the British government must establish itself in this field, offensively and defensively. Born in 1891 in rural Ontario, Banting was a medic during World War I, and, like others, he knew how unprepared the British and their allies had been to retaliate quickly against German chemical weapons. After being trained as a surgeon, Banting immediately switched to medical science, and his subsequent discovery of insulin saved millions of diabetics from death. In 1923, Bar shared a Nobel Prize with John James Macleod for this work.

Banting, a believer in the future importance of air warfare, envisioned the potential of airplanes at altitudes of 35,000-40,000 feet for dispersing bacteria over large areas. Although he knew little about bacteriology, he reasoned that infection could be imposed in a variety of ways. Conjecturing the use of air planes, he declared that "under the new conditions of bacteriological warfare, the disease producing bacteria may be scattered anywhere—and it must be remembered that as long as bacteria gain entrance to the body they produce disease."

Banting argued enthusiastically for the development of British biological warfare capacity. Without explicitly advocating the first use of germ weapons, he proposed that having an enormous retaliatory strength would give the United Kingdom a decided advantage over Germany. "If the Germans use bacteria he asserted, "the Allies should be in a position to retaliate one hundred fold without delay."

Banting set forth his ideas about retaliation, defense, and the German biological warfare threat in a position paper for the Canadian Defencc Minis try. He listed *tetanus*, *rabies*, *gas gangrene*, *parrot fever*, *anthrax*, *botulinum toxin*, and, against livestock, foot-and-mouth disease as possible destructive agents, and warned of insectborne diseases and assaults on water reservoirs. This paper was passed on to officials in the United Kingdom, but his perspective was rejected in favour of the public health approach, already developed by Hankey's science advisors. Edward Mellanby had presented this public health approach during a visit to Ottawa in September 1938 and it was the reigning government position.

In 1939, immediately following the German invasion of Poland and the British-French declaration of war, Banting, then forty-seven, rejoined the Canadian army as a major and devoted himself entirely to the war effort. While still head of his research institute, Banting chaired two high-level committees for mobilizing Canadian science. One was on medical research and the other on aviation medicine. For the latter, Banting invented the first program to investigate the effects of high altitude on pilots.

In November 1939, Banting traveled to England to lend his authority to the war effort. He and his scientific colleague, toxicologist Israel Rabinowitch, visited the Porton Down chemical weapons facility in Wiltshire and found its program in frill swing, although, hampered by the war, it lacked the space for large-scale field trials. Rabinowitch suggested that Canada's open spaces could be useful. The nearly one-thousand-square-mile. Suffield range in the province of Alberta later became a center for chemical and biological field tests.

Banting's more pressing mission on this trip was to persuade other influential scientists of the necessity to start a biological warfare program. He was convinced that Germany had been experimenting with bacteriological weapons for years, that it would not hesitate to use them, and the United Kingdom had to be prepared to retaliate by strategic attack. Banting took his case from one elite biologist to another with no results. On Christmas Eve of 1939, he had tea with Edward Mellanby, whose apparent lack of enthusiasm added to Banting's sense of discouragement. Banting also approached Maurice Hankey.

At this time, Banting was arguing that civilians were integral to modern war: "In the past, war was confined for the most part to men in uniform, but with increased mechanization of armies and the introduction of air forces, there is an increased dependence on the home country, and eight to ten people working at home are now required to keep one man in the fighting line. This state of affairs alters the complexion of war. It really amounts to one nation fighting another nation. This being so, it is just as effective to kill or disable ten unarmed workers at home as to put a soldier out of action, and if this can be done with less risk, then it would be advantageous to employ any mode of warfare to accomplish this." Having accepted this grim calculation, Banting put his mind to how a biological weapons program might be structured. In the memorandum he brought with him to London, he proposed a new organization.

This new organization would have three functions. The *first* would be togather and evaluate all known data on the feasibility of making *biological weapons agents*. For those skeptical of the stability of pathogens, Banting pointed out that organisms for *typhoid*, *whooping cough*, and *meningitis* had been successfully frozen and dried, and, by his reckoning, the problem of stability could be solved for other microbes. This part of the program would evaluate treatments to counteract these agents.

The *second* function of the program would be weapons research. Like others, Banting was interested in poison bullets. In addition, he advised experiments with shells and bombs and with the creation of a "*bacterial dust*" for dissemination from airplanes. He foresaw the need for factory-scale production of bacteria and a need for safety standards to protect workers. He theorized (correctly, as it turned out) that methods could be devised to bypass the vectors needed to transmit dangerous diseases such as plague and yellow fever.

The *third* part of Banting's proposed biological warfare program would be a "*Vigilance Committee*" a kind of intelligence wing that would track suspicious outbreaks and check up on possible espionage activities, such as the sale and use of laboratory equipment or other machinery indicating agent production. This committee was also charged with giving the public simple warnings, regulations, and instructions in the event of an outbreak; it might also, he suggested, consider the preparation of sufficient gauze nose masks for rapid, universal distribution if necessary to contain the spread of disease.

Banting, alert to the many ways an enemy might wage bacteriological war against the United Kingdom, even considered the possibility of bacteria being sent in envelopes through the mail, as indeed they were in 2001 in the US. He was especially concerned about water sources being poisoned. "Fixed ideas of natural epidemics must be altered in the face of artificial warfare epidemics," he warned. He ended his memo by offering, if the British government requested, trained Canadian bacteriologists, vast open spaces for testing, and the cooperation, through Canada's Department of National Defence, of artillery and air force units.

In February 1940 Banting left London convinced that his biological weap ons proposal had been rejected. He later dismissed Hankey as "a superb example of the servile, all-important complacent superior ass that runs the British government."

## "Fog of War"

Banting very much misjudged Hankey, who was impressed by Banting's argument for biological weapons research and its placement alongside the chemical program. Soon after Banting left England, Hankey asked his science advisory committee (reconstituted under the War Cabinet as the Biological Warfare Committee) to consider the Canadian scientist's biological warfare proposal. Although the committee as a whole remained unconvinced, Hankey ended the meeting by suggesting that experiments should at least be tried, preferably at Porton Down, the site of the established chemical weapons program. He held a private discussion with Mellanby and other more amenable committee members. Hankey then appointed John Ledingham and a subgroup to research the possibilities of germ weapons. Ledingham reported back to the Biological Warfare Committee that there were three likely uses of germ weapons. One was contaminated rifle bullets, which experiments showed worked. The second was the use of germ agents in sabotage, which the Germans had already tried in the previous war. The third was aerial dispersion, about which more needed to be known.

The committee resisted recommending any new government action. To them, poison bullets seemed too dangerous for workers to produce and for soldiers to use. Sabotage with microbes or poisons could work, but it hardly required an entire weapons program to kill a few adversaries with ricin or to infect horses with glanders or anthrax. And if bacteriological aerosols were feasible, the committee reasoned, they would require a considerable state enterprise to reach a dangerous scale of attack. The committee members could hardly imagine that Germany would divert its wartime resources to this effort. In this assessment they later proved accurate, for Germany, but not for their own government, Canada, and the United States.

On May 10, 1940 Winston Churchill replaced Neville Chamberlain as Prime Minister and also claimed the office of Minister of Defence.

The head lines that should have gone to Churchill's appointment went instead to the news that Germany had invaded the Netherlands. After years of service Maurice Hankey was now out of government. He and Churchill had previously clashed over a number of issues. Churchill though recognized Hankey's special organizational talents and wrote to Chamberlain suggesting that Hankey should become Chancellor of the Duchy of Lancaster—an appointment "in which he can continue the chairmanship of the many secret Committees over which he presides. In agreement Chamberlain wrote "V good" on Churchill's memo and underlined the notation twice to indicate his enthusiasm.

In July 1940 Hankey who then held no government office wrote in a memo, "I recently came to the conclusion that we ought to go a step further in the matter of bacteriological warfare so as to put ourselves in a position to retaliate if such abominable methods should be used against us." France had fallen to Germany the month before, and there were rumors that its secret biological weapons program was under German control. Germany was out to destroy British air power as a prelude to possible invasion and was bombing London and other cities to draw out the Royal Air Force. The Germans in 1940 had "the most effective short-to-medium range bombing fleet in the world and, unlike the RAF, German air forces had solved many navigational and bomb aiming problems." Alone and under violent assault, Britain was entering what Churchill called "its finest hour."

With German bombs falling, Hankey made the decision that the United Kingdom needed a biological weapons program and, as he often did, he acted on opportunity. The previous year, in early August 1939, authority over the Chemical Defence Experimental Station at Porton had passed from the War Office to the new Ministry of Supply. Without consulting with either the War Cabinet, which he judged far too busy to consider germ warfare, or Prime Minister Churchill, Hankey conferred privately with Sir Edward Mellanby and two other of his committee members about the problem of how to initiate a biological weapons program. He and Mellanby then went to the Minister of Supply, Herbert Morrison, to ask that biological weapons be developed at Porton Down, alongside the chemical program, which the Ministry oversaw. In this period of extreme national duress, the argument for new weapons made sense. Morrison agreed, and on August 18, 1940, he authorized a Biology Department to be created at Porton for biological weapons research and development. Hankey's rationale for sidestepping government committees, the War Cabinet, and the prime minister was that the biological weapons research was exploratory, like other experimental ventures undertaken within the chemical weapons program.

On the advice of John Ledingharn, Hankey recruited Paul Fildes, head of the Medical Research Council's Bacterial Chemistry Unit, to head the new secret department. A bacteriologist, Fildes took the position believing that his mission was to achieve high-volume offensive capacity available at short notice. As a civilian Fildes understood that he would answer only to Hankey, rather than to the military.

In early October, after arrangements with Porton had been made, Hankey had yet to inform the cabinet or the prime minister. He then received an offhand inquiry from Special Operations about possible crop destruction research. Hankey took this opportunity to write Churchill that the "*practicability*" of biological weapons had to be investigated in case it was ever used against the United Kingdom, at which point the military should be in a position "to retaliate if the Cabinet should so decide." Churchill approved. He had already pondered the future of biological weapons, writing in 1925: "Blight to destroy crops, *Anthrax* to slay horses and cattle, *Plague* to poison not armies but whole districts—such are the lines along which military science is remorselessly advancing."

Thereafter Hankey kept Churchill advised of Fildes's work, but Fildes's goal of offensive capacity (in the name of retaliation) still lacked formal approval from the War Cabinet. Finally, in January 1942, the cabinet informed Hankey of its approval, with ample warning that the authority for using such weapons rested exclusively with itself or the Defence Committee and that the program should be strictly secret. The next day a note was appended warning Hankey that the new program must not divert resources from the ongoing war effort.

To be quickly prepared for retaliation in kind, Fildes started with linseed cakes, which tests at Porton had shown cattle would eagerly consume. He then had five million made in a London soap factory and shipped to Porton, where local women, working in a small building nicknamed the "bun factory," injected them with anthrax spores. The project was first called *Operation Vegetarian*, then renamed *Operation Aladdin*. The cattle cakes, however, were never used.

It was calculated that a dozen Lancaster bombers in a single raid could drop enough anthrax cakes to threaten cattle throughout most of the north Ger man countryside. Had they been used, they might have caused massive animal deaths and subsequent human illness, in addition to reducing food supplies and damaging the leather industry. Since anthrax spores can persist in the soil for years, the cattle cakes could also have caused future outbreaks.

Fildes's main goal was to invent what Hankey's committee failed to imagine: a lethal anthrax aerosol widely dispersed by bombs. At Porton Down, experts in meteorology and physics, in munitions and in large-scale field trials over land and sea greatly assisted in achieving this goal. Fildes, an expert in growing bacteria, soon had a team working on a biological weapons model that approximated the one for chemical weapons. He recruited David Henderson, who was working on the chemical side of Porton and had done work on bacterial aerosols. By focusing on *anthrax*, with its tough spores and high virulence, Fildes had circumvented the fragile bacteria obstacle. With Henderson, he could investigate aerosols using anthrax spores. By the end of 1940, Fildes and his team had started small-scale spray and bomb tests at Porton.

## Biological Weapons Funding from Industrial Giants

In the summer of 1940, as Maurice Hankey was covertly initiating the Porton Down biological weapons program, Frederick Banting was taking advantage of an unusual opportunity that would shape the parallel Canadian effort and, indirectly, the US and UK programs. Following the fall of France, three wealthy Canadian captains of industry stepped forward with millions of dollars for war-related research—John David Eaton, founder of T. Eaton department stores; Sir Edward Beatty, president of the Canadian Pacific Railway; and Samuel Bronfman, the head of Seagrams. Banting was one of three scientists they consulted. The other two were engineer C. J. MacKenzie, head of the National Research Council, and Otto Maass, a well-known chemist committed to chemical weapons production.

Together, the six assembled a highly secret committee of science and military experts (the War Technical and Scientific Development Committee) to make decisions about innovative defense projects. Banting was soon awarded $25,000 to conduct laboratory experiments on germ weapons. Convinced of the necessity of both defensive and offensive biological weapons preparations, he organized around him a core of Canadian biologists and physicians to work on a wide variety of pathogens and on the physiological effects of chemical weapons. Banting's own laboratory at the University of Toronto was involved, along with the prestigious Connaught Laboratory, which was affiliated with the University and had a large facility ("the Farm") north of the city.

Banting's first biological weapons field experiment was to simulate an aerial attack. In early October 1940, near Balsam Lake northeast of Toronto, he succeeded in spraying sawdust from a low-flying plane. Afterward he and his assistants combed the ground to measure the dispersal path. The idea was not as far-fetched as it appeared. A few weeks later, one of Banting's junior scientists succeeded in drying typhoid bacteria on sawdust, a first step toward the mass use of a biological weapon against humans.

On the basis of his airplane experiment, Banting immediately went to the Secretary of Defense in Ottawa to argue for permission to produce bacteria for mass dissemination. Late that night, overwrought with patriotic feeling, he ranted in his diary about killing "three or four million young huns— without mercy—without feeling.... Those Huns at home, those huns of Hitler—It is our job to kill them."

Banting's ideas on biological warfare found favour with the Canadian Prime Minister, MacKenzie King, and on November 19, 1940, Banting received permission from Canada's Department of National Defence for bacteria production. Banting thought in terms of large-scale manufacture, a factory where a hundred tons of virulent organism could be produced with safety and speed. No such production plant was forthcoming in wartime Canada, but Banting's industrial-scale vision was integral to the US program and those that followed.

Through C. J. MacKenzie, word of Banting's secret enterprise traveled directly to Vannevar Bush, president of the Carnegie Institute and founder of the National Defense Research Committee. Bush, an advisor to President Frankjin Roosevelt, set in motion the first American inquiries into potential biological weapons strategies. Future contacts between US and Canadian scientists working on biological weapons proved direct and collegial. This was in contrast to the British presumption that Canadian Scientists were junior partners, mainly just supplying the wide open spaces for UK testing.

On February 21, 1941, en route to London and from there most likely to Porton Down, Banting was killed in an air crash in the Canadian Maritime Provinces. Banting's death caused mourning throughout Canada. A few days later, on February 24, Paul Fildes released his first report on successful *anthrax* aerosol tests and their possible biological warfare applications.

## Biological Weapons Experiments and Testing

Experiments and tests were vital to moving *biological weapons* technology toward assimilation by the military. Although *anthrax* proved tough and lethal in small tests, Fildes needed more precision in reckoning its virulence and infectivity. By April 1942, his team had perfected a cloud chamber for bacteria. The method of exposure was to hold the snouts of laboratory animals to apertures and let the aerosol pass by. *Cloud chamber* experiments facilitated determination of the aerosol particle size most effective for causing infection through the lungs and for estimating lethal doses for laboratory animals. For anthrax, a goal was to extrapolate from animal tests the amount of biological munitions and the number of aircraft that would be required to carry out a strategic aerial attack. Inhalational anthrax was known as a dangerous disease with no reliable therapy; to test it on humans would be impossible.

Fildes was eager to test the open-air effects of aerosols from anthrax bombs over a large area, on tethered sheep. Although relatively remote, Porton Down was too close to human habitation to pursue such potentially dangerous experiments. For this Fildes submitted a proposal to the Porton Executive Committee, which was then in charge of giving his projects oversight. Fildes had learned how to answer to government and, though never completely reconciled to military control over his research, he realized that his biological weapons needed "*potential users*," especially the Royal Air Force. Fildes received permission, and that summer a contingent from Porton's chemical mud biological programs traveled to Scotland and set up a base on the mainland across from Gruinard Island, where they would conduct historic bomb tests. Later during the war they would also conduct biological weapons tests at Penclawdd, on the coast of Wales.

It was in Fildes's favour that Prime Minister Churchill was open to unconventional weapons and already supported chemicals. Determined that the United Kingdom should have quick and massive capacity for chemical weapons retaliation, he pressed for increased production. The United Kingdom soon had twenty thousand tons of poison gas in storage, and 15 percent of its bomber force ostensibly at the ready to use it. Churchill also advertised this capability to the enemy. In an impassioned radio address on May 10, 1942, reacting to word that the Germans intended to use chemical weapons against the Soviet Union, now a Western ally, Churchill warned: "I wish now to make it plain that we shall treat the unprovoked use of poison gas against our Russian ally exactly as if it were used against ourselves, and if we are satisfied that this new outrage has been committed by Hitler we will use our great and growing air superiority in the West to carry gas warfare on the largest possible scale far and wide against military objectives in Germany. It is thus for Hitler to choose whether he wishes to add this additional horror to aerial warfare." This same fierce declaration inspired the American scientists Theodore Rosebury and Elkin Kabat to propose a retaliatory capacity for biological weapons in their 1942 report.

In July of 1942, as the trials at Gruinard Island were commencing, Churchill may have suspected that the Japanese, known to have chemical weapons, had developed biological weapons as well. At that time, the Chinese appealed to the West with accusations that the Japanese Imperial Army had spread plague against cities in eastern China. Although British colonial administrators dismissed the allegations, Churchill personally marked the report for circulation and discussion among his defense experts.

## First Field Tests

A formidable scientist from Porton, meteorologist Sir Oliver Graham Sutton, led the research team of fifty men to conduct the Gruinard trials. Henderson and other staff were in charge of the germ weapons, and Fildes was a frequent visitor. The team also included Chemical Defence personnel and military engineers, physicians, veterinarians, and munitions experts. They camped on the mainland, where rudimentary housing and a small laboratory were built. Gruinard was an uninhabited island about two kilometers long and one kilometer wide, covered with grass, heather, and bracken. About 155 sheep bought from local villagers were used in the tests.

The first trial at Gruinard, on July15, used a modified thirty-pound bomb filled with three liters of anthrax suspension. Instead of risking the imprecision of an airplane drop, the bombs were suspended on gallows four feet above the ground and then detonated by remote control, producing a visible cloud that drifted downwind. After seven days, nearly all the sheep that had been tethered in the range of ninety to one hundred yards from the source were dead. Further tests showed that the anthrax cloud was lethal at up to four hundred yards. For a more realistic scenario, in September a Wellington bomber dropped an anthrax bomb from seven thousand feet; but the bomb landed in a peat bog and had no effect on the sheep. The team turned to another site, one previously used by *Chemical Defence*, at Penclawdd, in a beach area. There a bomb dropped from a plane at around five thousand feet missed its mark by only twenty feet. Sheep in the range of 120 to 320 yards from the explosion died soon after from anthrax. Trials continued at Gruinard throughout July and September 1943 and their results were reinforced by experiments with simulants at Porton.

The Gruinard tests suggested that, weight for weight, *anthrax* could be one hundred to one thousand times more potent than any chemical weapon of the time, to a distance of five hundred yards. Further, an efficient electrically fired four-pound bomb was developed; 106 of them could be packed into a cluster bomb derived from models used for incendiaries in World War I. Fildes had also defined the approximate lethal dose for sheep, an important standard for further research and production. These tests, only the first of many to follow, also began training soldiers through simulated biological warfare.

In August 1943, the tests at Gruinard were forced to a halt. The previous spring, after a heavy storm, dead sheep buried on the island near the coast apparently were dislodged and floated to the mainland, where they caused more than fifty deaths among livestock. The government told the alarmed villagers that a nearby Greek ship on maneuvers had tossed some infected carcasses overboard and compensated them for their lost animals. Wary of more accidents, Fildes hoped that future tests could be conducted in the open spaces of Canada and the United States.

Fildes's claim to the first modern biological weapon circulated through the various committees charged with national defense. The favourable comparison of anthrax to much heavier chemical agents meant that biological bombs might have a future. Fildes's success was taken as proof that the Germans must have developed a similar biological weapon, although there was no intelligence information to support this conclusion. Fildes himself firmly believed that Germany had an identical weapon and he continued to argue that the power of retaliation was the United Kingdom's best defense against the threat. His technical breakthrough, bombs that killed sheep, became the "*looking-glass*" justification for a full and ready offensive biological warfare arsenal.

As he succeeded in his research, Fildes defended biological weapons with the humane argument that during World War I the Germans had used for chemical weapons, and which the noted British geneticist J. B. S. Haldane had also argued. Those arguments referred to soldiers. Fildes saw his efforts as benefiting civilians as well. In response to a criticism of biological warfare as "*unsoldierly*" and eliciting "*public revulsion*," Fildes wrote: "Is it any more moral to kill Service men or civilians with HE (High Explosives) than with BW? . . . It seems clear to me that a substantial majority of the population would conclude that, if they had to put up with a war again, they would prefer to face the risks of attack by bacteria than bombardments by HE." From Fildes's point of view, no moral or technical obstacles prevented the next logical step, the large-scale production of anthrax bacteria and thereafter the mass production of anthrax bombs. From a strategic perspective, the *anthrax bomb* opened up new potentials for destruction, uncoupled from retaliation.

Fildes was constantly in communication with the Canadians and Americans about the successes of the *anthrax bomb*, which he wanted mass-produced. Britain was deeply mortgaging itself to pay for the war, and he had been warned that biological warfare projects must not burden this effort. Canada had neither the skilled labour force nor the other industrial resources to spare. Fildes had taken the United Kingdom as far as it could go as with biological weapons. For industrial manufacture, his project needed the United States.

# 2

# BIOLOGICAL THREAT AND ATOMIC BOMB

The American officers who interviewed Japanese biological warfare scientists in 1947 quickly realized that US achievements were technically superior to those of Gen. Ishii Shiro's program. By the end of World War II, the American, British, and Canadian programs had learned much about airborne infection and about a diversity of biological agents, had im proved defenses against some of them, and had come close to mass production of an *anthrax bomb*. The atomic bomb and the Cold War signaled a momentous change in the US biological weapons program. The vision of the scale of intentionally spreading disease expanded to strategic attacks on a par with the destruction of Hiroshima and Nagasaki and with the Soviet Union and its allies as potential targets. For more than twenty years after the end of World War II, advocates of germ weapons struggled toward the goal of strategic potential and sought to convince the armed forces, especially the Air Force and Navy, that these new weapons should be assimilated. Few in the US government knew about this struggle for expanded offensive capacity. Within the program, research intensified, production capacity increased, and simulations of disease attacks on civilians grew larger and more elaborate until they verged on reality.

During the time the US postwar biological weapons program was being secretly ratcheted up to approximate nuclear scale, members of the American and international public, including scientists, were openly debating the dangers of nuclear weapons. Eventually, especially during the 1960s, the US chemical weapons program and the dangers it posed to civilians and the environment became a source of great contention. Yet, after just a brief flurry of publicity about US wartime research, biological weapons disappeared from public discussion for two decades. In 1947, Dwight Eisenhower, then Army Chief of Staff, banned disclosures regarding the program's continued activities. Following this, in a 1949 press statement, Secretary of Defense James Forrestal emphasized only US defensive research on biological weapons. His point, which the public seemed to accept, was that the biological program existed to defend the nation against the Soviet threat. What the public did not realize was that this program for defense was squarely based on offensive goals. Since the United States was not bound by the Geneva Protocol, military advocates for biological and chemical weapons set about campaigning for the right of first use, which would allow the development of a fully offensive biological weapons program.

A 1951 report from the office of the Joint Chiefs of Staff (JCS) summarized the perceived military potential of biological warfare, including strategic and sabotage uses. First, the report noted that the relative cheapness of *biological weapons*, compared to conventional, chemical, and nuclear arms, was a major benefit. The concern that lesser nations might acquire these weapons, which became central later in the century, was lacking; instead the presumption was that only industrial powers had the resources for biological weapons, as they did for nuclear and, in general, for chemicals. Thus, the JCS report concentrated on the USSR as an adversary and presented a number of World War II—type scenarios drawn from both the European and Asian theaters. In one scenario, the so-called *nonlethal diseases* (such as *brucellosis*) were suggested as a follow-up to atomic bombing in order to further debilitate enemy civilians in the area of attack. The use of biological agents in special operations was especially recommended, with civilians as unsuspecting targets: "Probably one of the most attractive and profitable uses to which BW could be placed would be in the field of covert operations. The ability of special agents or guerrilla forces operating behind enemy lines to place small quantities of BW agents accurately where they could be most effective offers tremendous possibilities. Covert use has a further advantage in that distinction between this type of BW and naturally occurring outbreaks is most difficult, and therefore might be used with crippling results well in advance of the initiation of hostilities."

On a technical level, the first demand of the military was that this new kind of weapon must be dependably destructive. Commanders had only one goal, to win wars, and they were reluctant to experiment with germ weapons in battle. A 1952 JCS memo expressed this essential requirement and went further to indicate two important directives. One was that the Army's Chemical Corps should conduct field tests of biological weapons and the other was that each of the three services should prepare to accept biological weapons through the education or "*indoctrination*" of personnel:

The Joint Chiefs of Staff desire to Stress the fact that, regardless of the potential value in various situations of the available BW weapons, these will not be used unless the operational commander has confidence in the superior ability of the weapon concerned to contribute to the success of his mission. Therefore, tests to determine the operational capabilities of BW systems, and indoctrination measures to disseminate and create confidence in the test results, are necessary. Subject to coordination by the Secretary of the Army, who has been assigned responsibility for coordination of the field test program, the three Services [Army Naval Operations, and Air Force] should conduct field tests on a scale sufficient to determine the military worth of candidate agent munition combinations, their offensive uses and the means of defense against them. In addition, indoctrination in BW through the use of orientation teams, courses of instruction, on-the-job training, or by other appropriate means should be undertaken in each Service, with such mutual assistance and cooperation as may be practicable.

Despite intensive development and testing, throughout this era biological weapons were neither assimilated into the thinking and planning of the regular military nor used by the United States or its partners in this endeavor—the United Kingdom and Canada and, later, Australia.

Two wars presented opportunities, the Korean War (1950-1953) and the Vietnam War (1962-1975). The Korean conflict involved the United States as the leader of a multilateral, United Nations effort that would have dampened any impulse to use biological weapons. In addition, the US program, still struggling to develop more efficient biological bombs for the dispersal of *anthrax spores* and other pathogens, had little reliable to offer. Indeed, in 1952 China and North Korea accused the United States of airplane attacks with contaminated insects, feathers, and other vectors, at the technical level of the Japanese program.

During the Vietnam War, the national and international controversy over US use of chemicals in combat, especially riot control agents, suggested that the introduction of ***biological weapons***, covertly or not, would have caused a furor. During the 1960s, field tests showed increasing promise. Those involving tropical diseases and designed to penetrate jungle canopies might have tempted some military planners frustrated by Vietcong guerrilla tactics and wanting every attack option, but US government policy held the line.

## Biological Warfare, the Atomic Bomb, and the Soviet Threat

In May 1946, based on a leak from a closed hearing on the Navy budget, the Associated Press carried a sensational story about a germ weapon "far more deadly than the *atomic bomb*" and "capable of wiping out large cities and entire crops in a single blow." AP withdrew the story within hours, stating that no such weapon existed. In this and other ways, the public was prepared to perceive biological weapons as a new order of threat. The US press was then discovering the old wartime biological weapcns program, with help from the Army and the Navy that emphasized the defensive measures that the US military had developed. The Navy, for example, held a press conference reporting that its research unit in Oakland, California, had used fifty San Quentin prisoners to study bubonic plague as a means of developing medical defenses. The New York Times, the Washington Times-Herald, Newsweek, Life, The Nation, Science Illustrated, and a score of other publications picked up on the postwar research publications from Camp Detrick and ran stories on the frightening strategic potential of biological weapons. A headline in Science News Letter read, "Germ Warfare Advances More Potent Than A-Bomb."

In 1947, as the immunity bargain with Japanese scientists was being negotiated and as the Cold War commenced, the United States submitted a draft resolution to the United Nations that defined weapons of mass destruction as including "*atomic explosives*, *radioactive material*, *lethal chemical* and *biological weapons*, and any weapons developed in the future which have characteristics comparable in destructive effect to those of the atomic bomb or other weapons mentioned above. This grouping of unproven biological weapons with chemical and nuclear weapons continued through subsequent administrations, Democratic and Republican.

With the *Cold War* escalating, the obvious question within government was whether the USSR posed a biological weapons threat to which the West, particularly the United States, should respond by having an offensive capability. US military and intelligence assessments acknowledged the devastating effects of the war on the Soviet economy and on health of Soviet civilians. There were known gaps between Western and Soviet technology—especially with regard to the manufacture of antibiotics and industrial fermentation capacity. Yet the Soviet Union had the scientific means and the intent to pursue nuclear weapons; might it also aspire to biological weapons? Suspicions of a Soviet biological weapons program became common in Washington. In 1948, for instance, US Senator Edwin Johnson told the press that he had it on good authority that "the Russians have perfected this terrible warfare weapon of spreading plagues and germs."

After the Soviet Union tested its first atomic bomb in August 1949, US military analysts speculated that biological weapons would be its next strategic priority. By 1951, although intelligence on the subject was poor, a JCS report concluded that the USSR had biological weapons and was using human subjects to test them in field trials: "Soviet interest in the BW field is of long standing. Unconfirmed reports indicate that considerable effort has been exerted and that this effort is continuing. It must be assumed, and it would be foolhardy to do otherwise, that the Soviets have developed an offensive capability in the field of biological warfare. We can be sure that there is no hesitancy on the part of the Soviets to obtain factual information on the effectiveness of their own BW agents and munitions by large-scale tests using human subjects."

## Biological Weapons Startup

In 1946, the annual US Chemical Warfare Service (CWS) budget was reduced to $933,000, down from its peak of $2.4 million during the war. In 1947, as the Cold War began, its allocation returned to its wartime level, $2.75 million. The postwar discovery of German nerve gas, unknown in the West, had moved the CWS in a new direction of research. Biological weapons were an other reason for increased funding. Under Alden Wait, commander of the chemical program, those in charge of biological weapons streamlined their research, concentrating on intensive studies of known pathogens and toxins and their adaptation to bombs and aerosol generators. Qualified young scientists were recruited to the program, helped by fellowships from the National Research Council of the Natural Academy of Sciences. Available records indicate that the postwar program went through difficult periods, due mostly to technical problems, until it reached relative stability in 1954.

The Army's postwar biological warfare research, development, and pilot-scale production program was based at Camp (later Fort) Detrick in Mary land, with additional facilities at the animal research station at Plum Island (Fort Terry), New York. Detrick also participated in projects at the Navy's Oakland Biological Laboratory in California and in various covert intelligence projects. Field testing was centered at Dugway Proving Ground in Utah. The program was tasked with procuring empty bombs from civilian contractors, filling them with biological agents or innocuous simulant payloads, and transporting them to test sites. The biological weapons program also used a variety of environmental sites in the United States and elsewhere for open-air testing, drawing on its partnership with Canada for the use of sites there and cooperating with the United Kingdom in its field tests. When the Chemical Corps requested a new facility for biological agent and munitions production to replace the wartime Vigo plant, Congress responded with funding for Plant X-201 at Pine Bluff, Arkansas. The ten-story facility, started in 1950 and completed in 1953, cost $90 million and eventually employed 1,700 workers.

The United Kingdom's program at Porton Down seems to have made a smoother transition to strategic objectives and expanded field trials than the US program. David Henderson, Paul Fildes's second-in-command, took over as director, and the program maintained knowledgeable advisors, such as Paul Fildes, Lord Stamp, and Maurice Hankey, who also offered continuity. The British were against revelations about their wartime activities, which nonetheless did attract unwelcome press attention. As in the United States, the UK program was drawn to the development of biological weapons on a par with atomic ones. Having ratified the Geneva Protocol, its struggles to resist fully offensive objectives were more evident than in the United States, and at various times its program leadership claimed a purely defensive mission or appeared to limit itself to cooperation with special operations. If biological weapons had a future, though, it was for effective, large-area attacks. The postwar program at Porton Down, therefore, was committed to the same two goals as the American one: a vigorous laboratory investigation of potential agents and the initiation of large field tests with pathogens, toxins, or their simulants.

In 1945, the Biology Department at Porton was renamed the Microbiology Research Department (MRD). In 1948, despite the United Kingdom's dam aged economy and a shortage of construction materials, ground was broken at Porton for an immense new facility, the largest brick building in Europe at the time. Henderson felt that understanding inhalatory infection should be the focus of laboratory research, and he arranged for the building of a bomb-bursting chamber for expanded aerosol testing. Soon after, at Camp Detrick, the Americans built the forty-foot-high, million-liter Horton Test Sphere, the largest in the world. The Horton Sphere allowed explosive munitions to be detonated inside the chamber, with caged test animals exposed through portals.

Finding scientists to work at Porton was difficult; the responses to job advertisements were disappointingly low. Scientific advisors took a dim view of postwar media commentaries they felt turned the public and perhaps prospective staff against the program.' In any event, the Porton laboratories, designed for 130 scientific researchers and technicians, employed only 72. The program did, though, attract quality researchers who contributed to defenses against biological weapons. For example, biochemist Harry Smith, who arrived at Porton in 1948, discovered the three toxins produced by anthrax bacteria. This discovery became the basis for the creation of the antigenic anthrax vaccine, still in use a half-century later, and for modern antitoxins.

Large-scale production was also on the Porton agenda. The MRD was funded for another large facility and two satellite pilot fermentation plants. Should a war break out with the Soviet Union, it was reckoned, Porton should be ready with the necessary production rate of thirty-five tons of agent per day, to supply the four-ton bomb capacity for each of five hundred aircraft conducting four sorties per month. Restraints from the British Cabinet and elsewhere among civilian leaders in government soon undermined this goal by imposing restrictions on building.

## Technical Restraints

As early as November 1945, the UK Chiefs of Staff were speculating about the future of biological weapons in ten years' time, compared to that of nuclear weapons. A War Office memo voiced optimism about biological weapons, noting that, with more support, the program might produce a bomb that could cause a 50 percent risk of death in an urban area of three square miles.

In 1946, the UK Chiefs of Staff were committed to preparing the nation for the use of weapons of mass destruction, which they identified as nuclear, chemical, and biological. Under an Air Staff research project called "*Red Admiral*," Porton was to assist in creating an antipersonnel bomb by 1955. The Air Staff reported that despite its aversion to initiating a biological attack, as last resort capacity, it wanted a thousand-pound cluster bomb, a "*parent*" bomb containing many small "*child*" bombs, for wide-area coverage. Its estimate, following from the previous year, was that two hundred tons of this type of bomb could cause 50 percent death rates in an industrial city of one hundred square miles. High-altitude bombing was preferred, for which Porton would have to develop accurate meteorological reports and protect the disease agents from extreme cold. The report also considered spray generators, if researchers could improve their targeting.

Neither the mechanical effectiveness of biological bombs in generating respirable aerosols nor the dose response for such aerosols was sufficiently tested to allow accurate estimates. Persistent technical problems had troubled the modified anthrax bomb during the late years of the war and continued to do so. For dangerous pathogens without reliable antibiotic therapy, scientists were limited to animal experiments from which they attempted to extrapolate the human dose response, that is, what inhaled dose of an agent might infect what proportion of a target population.

The US Air Force was initially enthusiastic about the potential of biological weapons, but its enthusiasm soon wore thin, as did that of the British Air Staff. By 1950, although flush with funding, biological-warfare scientists still had not perfected the four-pound Mark I prototype bomb and so decided to shift priority to a two-pound munition for cluster bombs. In February 1951, the JCS placed biological weapons in Strategic Category and charged the Air Force with developing a worldwide combat and defensive capacity. The biological weapons program still had not produced an efficient bomb. This failure was a sharp disappointment for the Air Force, which had placed procurement orders before waiting for bomb tests. The program's other invention, the four-pound $M_{33}$ bomb, had only small area coverage, and its carriage design was already obsolete for new high-performance bombers. "Research and development had failed to support planning estimates' Dorothy Miller, an official historian of the Air Force biological program, wrote, "and the near future held out small hope for substantial improvement."

The Air Force staffs involved with biological warfare were organizationally divided; one was concerned with research and development and the other with logistics and training. The first office was unsure of the technical future of biological weapons. Logistics and training had also lagged. Few military personnel knew how to handle or transport biological weapons or what to do in the case of accidents, or even how the bombs would be transported from the Chemical Corps for secure loading at airfields, or who would be in charge once the munitions went overseas. As Miller observed, "Logistical support of munitions was a nightmare." A plan to train a hundred medical and science officers in biological warfare fizzled; the Chemical Corps had little to tell the Air Force that was not speculation.

## Korean War Accusations

In this unsettled period, allegations of US use of biological weapons in the Korean War became an international issue, debated at the United Nations. In early 1952, the Chinese and North Korean governments accused the United States of using biological warfare tactics learned from the Japanese.

Rejecting the International Red Cross and the World Health Organization as biased toward the West, the North Koreans and the Chinese appealed to the World Peace Council to form an investigatory committee. This international committee of scientists left in June 1952 and spent two months in China and North Korea evaluating evidence for the allegation. The committee interviewed eyewitnesses and physicians, as well as four American captives (a pilot and three navigators) who had testified to US use of biological agents. The leader of the committee was the eminent British chemist and historian of Chinese technology, Joseph Needham, who had been stationed in China during World War II. Soviet biologist N. N. Zhukov-Verezhnikov, who had been the chief medical expert at the 1949 Khabarovsk trial, was also a member. Although it conducted no research of its own, the committee concluded that the United States had conducted biological warfare, primarily by airdrops of agents or contaminated materials. It compiled a seven-hundred-page report detailing fifty alleged incidents. The report included autopsy reports and testimony by physicians who had treated patients and scientists who had analyzed data, along with photostatic copies of supporting statements handwritten by the captured US airmen. The Needham report, as it is often called, contains allusions to the Japanese use of disease agents in China and includes claims that the US government was continuing to protect Japan from its criminal past. The report also cited postwar published articles by US biological warfare scientists as proof of an American offensive program and reminded readers that the United States had failed to ratify the Geneva Protocol.

The US government denied the charges of biological weapons use, and this denial was generally supported in the West and by eminent scientists who challenged the Needham report. The allegations are still a matter for study. Recently, for example, a Japanese journalist claimed to have seen Soviet documents indicating that the evidence of the Korean War germ attacks had been faked. The controversy demonstrated the difficulty of separating evidence from ideology in the case of suspicious disease outbreaks and also reinforced the dictum that intelligence in time of war is at best faulty. With ideological adversaries locked in battle, no credible independent investigation, allowing scientists to pose critical questions and collect their own data, seemed possible.

## Toward Improved Testing

Meanwhile, within the secret US program, scientists were having problems producing evidence that their biological weapons could be effective. Either the biological bomb tests were inconclusive or their more promising out comes were poorly communicated to Air Force personnel, who remained unpersuaded. Specifically, Air Force officials had grown impatient with the biological program's reliance on animal data. They wanted medical accuracy that would allow better bombs. Yet human-subjects research using biological agents could only be conducted if quick and reliable postexposure therapies were available, or else the research was ethically impossible.

In April 1953, recommendations to cancel all procurement agreements with the biological weapons program were circulating through the Air Force, at the level of the Secretary. One memo referred to the program as "*an unwarranted luxury*." Rather than gaining confidence in biological weapons, Air Force staff perceived the program as an unstable environment, with support liable to "*booms and busts*." Compared with the larger conventional and nuclear programs, it had limited career value for talented officers.

Air Force confidence in biological weapons was at an ebb, but the Army continued its support. On March 1954, a new directive from the Defense Department limited the program to long-range research and development, coupled with field testing. The biological program would research only improved munitions and concentrate on "*lethal* anti-personnel biological munitions for release from high-speed aircraft [one-half pound E61 anthrax bomb] and upon the development of munitions for delivery of anti-crop pathogens from high speed aircraft and by balloons." An "*acute disillusionment*" experienced by former enthusiasts translated into budget cuts. The Air Force investment in biological weapons research and development, at a high of million in 1953 fell to $1 million in 1956. Other Air Force categories, including the purchase of antipersonnel munitions from CWS and industry, showed similar declines.

The biological weapons program had two ways of increasing confidence in its weapons, and its scientists used both of them. The first effort was to improve field tests, which included secret sea and land trials with real pathogens and with simulants over UK, US, and Canadian urban targets. Improved estimates of aerosol dispersion patterns and agent survivability and virulence in various settings, under various weather conditions, relied on these data. As they grew larger, the field trials also were a way of gaining operational experience relevant to the conduct of strategic biological attack, uncoupled from the real-world contingencies of causing disease outbreaks and the radical consequences for war and society.

The second way to improve confidence in estimates of biological weapons effects was to embark on human-subjects research with pathogenic aerosols. After the war, these experiments were made possible through the availability of antibiotics for postexposure treatment, particularly for curing tularemia. The calculation of dose response was critical. It had to be known at least approximately in order to establish requirements for munitions performance and to estimate how many munitions would be required to attack any particular target. For its part, the Air Force needed to know how many bombs would create a desired target effect so it could plan how many planes would be needed. The UK and US open-air field trials with simulants and with real agents were tests of the feasibility of this grand plan.

## Operation Harness

As with much biological warfare research, the British were first with both sea trials and the investigation of the dispersal of biological agents over urban areas. "Operation Harness' the 1948—1949 British series of field trials off the Caribbean islands of Antigua and St. Kitts, was the first in a limited series sponsored by the United Kingdom. The United States followed suit and continued sea trials for years. The choice to hold the trials at sea was determined by fears of soil contamination (especially with anthrax, which had badly contaminated Gruinard Island), of public disclosure, and of possible danger to local populations. Unfortunately, high winds in the Caribbean forced the testing closer to land, where they were conducted "with many small fishing craft in the vicinity."

American scientists cooperated in Operation Harness, supplying sonic pathogens and bombs (the four-pound American E48. R2 type), along with bursters and modified fuses. The bombs were filled onsite and then suspended fuse downward for detonation twelve to eighteen inches above the ocean surface, for dispersal over caged animals on dinghies. The test animals included guinea pigs, sheep, and monkeys imported from India; it was the first time that guinea pigs and monkeys were used in *anthrax aerosol tests*, and both would frequently be used in subsequent research. Over the course of about two months, twenty-seven trials with bacteria, including some dispersed by spray generators, were carried out. David Henderson later reported that major miscalculations had been made, that the project had been rushed and its results were few. Although he believed the sea trials could be improved if plans were better, he also argued for better scientific investigation of aerosols, even referring to field trials as a "*blunderbuss method*" to gain knowledge about dispersal processes. The summary judgment from UK authorities was that more such trials should be conducted, and some were. It was the United States, though, that embarked on dozens more such tests of increasing sophistication and elaboration, and that turned its attention to experimental aerial tests in urban areas, as the British had also tried, and from here to large regional simulations.

## St Jo Program

In the United States, the Air Force wanted "a methodology for predicting weapons effects and for estimating munitions expenditures. To acquire this methodology it needed to understand the predictable dispersal of aerosol clouds over the potential target areas." The request was reasonable. By 1951, the *Chemical Corps* intended to give the investigation of aerosols top prior ity, and expenditures of over $4 million were expected for the period 1950-54. The technical goal was uniform dissemination of agents of around one to seven microns in particle size, the optimal size range for human inhalation, which could lead, for example, to *anthrax infection*. The postwar program had kept its ties to industry; several types of spray generators were being developed by independent contractors.

In response to the Air Force demand, Detrick scientists put together the St Jo Munitions Expenditure Panel. Its mission, begun in 1953, was to simulate anthrax attacks on urban targets. Three North American cities were chosen to approximate Soviet cities: St. Louis, Minneapolis, and Winnipeg. From January to September, its experimenters used generators mounted atop automobiles parked in various urban locations to disperse clouds of simulants. The presumed conditions of the potential attack were stated clearly: "Any expenditure figures finally derived will refer to a completely unprotected target population, which is assumed to be exposed in the open in a city, during the whole time of passage of the biological cloud."

In the late 1940s, the British had already tested an urban target by dispersing simulants on the city of Salisbury, near Porton, after which they carefully measured the influence of meteorology on cloud dispersion. In 1950, US ships had secretly sprayed aerosols of the anthrax simulant *Bacillus globigii* (BG) and *Serratia marcescens* (the tracking germ referred to in the 1934 Wickham Steed report) to be carried by wind over the San Francisco Bay Area.

The St Jo program brought a new array of science to bear on offensive simulations. Its first trials, which ran intermittently from January until September 1953, integrated every data source Detrick experts had: munitions testing and animal dose-response studies at Detrick, open-air trials at Dugway using BG, and meteorological studies commissioned jointly from Stanford University experts and a private contractor. The scientists in charge used data from the UK Salisbury tests and from earlier CWS experiments with chemical dispersion. The program also enlisted the help of computer programmers from the National Bureau of Standards.

In August 1954, the St Jo panel published pages of methodology, with maps of aerosol dispersal patterns from urban trials and dozens of closely argued mathematically formulas, concerning predicted doses of anthrax and the expenditure of bombs it would take to produce them. From the Air Force perspective, the St Jo program "represented a major advance in the concept for testing a biological munition-agent combination. Most important, it showed how, depending on meteorology and terrain, the number and pattern of bombs dropped on a target were related to the percentage of "*casualty production*" from anthrax aerosols.

The panel for the St Jo program tempered its report with scientific skepticism and caveats. For example, it reckoned that eight thousand anthrax spores would infect 50 percent of humans exposed (the lethal dose for half the population, or $LD_{50}$ but cautioned that this figure was only an extrapolation from monkey research data. It noted that the E61R4 bomb had limited aerosol efficiency. An effective aerosol would deliver anthrax particles in the one- to seven-micron range;

only 2 percent of the aerosol from the E61R4 was in this range. Adding more bombs or increasing the virulence of the agent (which had been done for anthrax) could increase casualties.

St Jo gave needed new hope to the biological weapons program by pointing to the potential of aerosols and bombs, if only the technical obstacles could be overcome. The Air Force consequently became interested in those technicalities—what the program could do, for example, to increase an agent's virulence or improve aerosols by using dry agents instead of slurries. In another development, it asked for more study of less lethal "*incapacitating*" diseases, such as *brucellosis*, and of highly *contagious diseases*, including *smallpox*.

Field trials remained essential to the program, and their numbers escalated. From 1950 to 1954, seventy-nine trials of BG were staged, including sea trials off the coasts of Virginia, California, and Florida. BG land trials took place at Dugway Proving Ground; Camp Detrick; Elgin Air Force Base; Fort McClellan, Alabama; Fort Belvoir, Virginia; Wright-Patterson Air Force Base; Loring Air Force Base, Maine; and Kirtland Air Force Base, New Mexico. In 1955, two dispersion tests were conducted on major state highways in Pennsylvania. From 1951 to 1955, at Dugway, there were twenty-five open-air field trials with pathogenic agents, including those for *anthrax*, *plague*, *brucellosis*, *Q fever*, and *tularemia*. Some of these trials, for example, a 1954 Banthracis trial, ran for months. The use of fluorescent particles to test wind dispersal in public places increased greatly once the Stanford University meteorological test results were reported. In addition to the tests in Minneapolis and St. Louis for the St Jo program and the Bay Area tests, open-air trials were con ducted at Corpus Christi, Texas; in the Mojave Desert; and, in 1957 and 1958, in unspecified areas east of the Rocky Mountains. In 1959 and 1960, thirteen trials were conducted in unspecified locations in north central Texas. This activity would increase sharply as the program entered the 1960s.

## Project Whitecoat

A key phrase in the St Jo methodology report was the disclaimer about the lethal dose of anthrax in humans, that the dose response could not be proven without actually exposing human subjects to anthrax aerosol, which would be highly dangerous. Because, unlike anthrax, the bacterial agent of tularemia (*Francisella tularensis*) could be rapidly and reliably cured by early administration of antibiotics, it held promise for human research. Although not as lethal as inhalational anthrax, *tularemia* (also known as *rabbit fever*) could cause debilitating disease and death rates of as high as 50 percent, depending on the strain.

In 1952, at the time when the United States suspected the Soviets of conducting unethical field trials using human subjects, physicians delegated to Camp Detrick by the Surgeon General's Office had the idea to start tularemia research. Their first attempt to use enlisted men at the base caused a sit-down strike. Then they considered research on Seventh Day Adventists who were conscientious objectors. Starting in 1953, after an informal agreement with leaders of the *Seventh Day Adventist* (SDA) Church, the medical corps began a long series of weapons tests on church volunteers. The pilot phase was called CD-22 and involved ninety-one volunteers' being exposed to aerosols of Q fever, which could also be reliably cured by *antibiotics*. In 1954, on the basis of the success of CD-22, Camp Detrick moved forward with Project White coat.

Twice a year a US Army officer and a church elder traveled to Fort Sam Houston, Texas, where conscientious objectors were trained as medics. There they recruited only Seventh Day Adventists for the Whitecoat project. Around 2,200 SDA men volunteered in the course of the program. From 1955 to 1957, trials were based on injections of *Pasteurella tularensis*. In September 1957, a project officer began a series of "*respiratory challenges*" that is, aerosol exposure in order to gauge dose response. Some volunteers were sick for days or weeks despite antibiotic therapy. Others recovered quickly. Tests were also conducted on how, once infected at various doses, the men were able to perform simple tasks or do mathematical calculations.

Nearly two hundred human-subjects experiments were conducted under Project Whitecoat. The peak years were 1963 through 1966, when most of the experiments involved tularemia, *Venezuelan equine encephalitis*, or *cutaneous anthrax*, a form for which antibiotics were fully effective, in contrast to anthrax acquired by inhalation. In 1967 through 1969, as the program was ending, a switch was made to studying tropical diseases, primarily sand fly fever and yellow fever, but also Rift Valley fever, plague, and Chikungunya, a mosquito-borne virus. Project White coat ended with the 1973 abolishment of the draft, although Army history shows that some tests continued until 1976.

The Project White coat participants were physically well treated and seemed, from Army checkups over the years, to suffer no long-term health consequences. The problematic issue was their ignorance of the goal of the research. Church leaders had understood that the project was for defensive purposes, to discover maximum protection against tularemia. Yet the main goal of the project was offensive, to ascertain human dose response for tularemia and thereby further the development of an effective bomb and, from there, the military assimilation of biological weapons for strategic attacks.

Industry was another source of human data on *biological weapons*. In the 1950s, in order to conduct control studies on its anthrax vaccine, the Army sent scientists to woolen and textile mills and other factory settings known for high rates of anthrax spore contamination. In 1957, in a felt-making factory in Manchester, New Hampshire, five unvaccinated workers came down with inhalation anthrax, probably from a highly contaminated shipment of imported animal hair or wool, and four died. The investigatory work on those deaths, including autopsies, added to the Japanese data. As occupational health standards improved and as the textile industries grew outmoded or went overseas, the factory environment for studying anthrax disappeared.

## Options for First Strike

As the biological weapons program struggled for stability, a major policy change took place that would free the biological and chemical programs from the constraints implied by earlier US tacit acceptance of the Geneva Protocol, short of ratification, and bring decision making for both in line with nuclear weapons policy. At the time, US nuclear policy left the ultimate decision for use with the president but allowed considerable latitude for the Defense Department to plan for possible first strikes. Nuclear deterrence played a part in this policy. Deterrence had also played a part in restraining the use of chemical weapons in World War II. The role of deterrence regarding biological weapons was much more ambiguous. Rather than deter the Soviet Union by threatening it with disease weapons, the United States was proceeding in secret to develop a new strategic weapon. Advocates of this venture within the military and the few members of Congress who approved the CWS budget wanted no restraints on the program.

In 1950, US policy for biological weapons was retaliation-only, in line with the Geneva Protocol. This policy was described as "*interim*," meaning that it could change as weapons were improved and reevaluated. In May 1954, the Army Chief of Staff called for a review of retaliation-only use of biological and chemical weapons. As Miller recorded, the argument about this restraint was longstanding. The Army felt that "a major obstacle to BW progress was the reluctance of the military to invest time and resources in a weapons system that might never be used." With a first-use option, a full development and military assimilation could take place: the United States could pick the time, place, and terms of an attack instead of merely reacting. And it would not be limited to attacks against the Soviet Union; it could attack any enemy. Critics within the military argued that germ weapons were simply not competitive with conventional and nuclear weapons and that only the requirement for a retaliatory capability against the Soviet Union kept the program alive.

The decision about a policy change was ultimately passed up to high-level security advisors who decided to free the Army from the retaliation-only use of both biological and chemical weapons. The justification cited was a new intimation of a Soviet biological threat. The source of concern was the February 20 speech of Soviet Marshall Georgiy Zhukov, the same military leader whose troops had been the target of Japanese biological weapons in 1939. At the twentieth Communist Party of the Soviet Union Congress, Zhukov declared, "Future war, if they unleash it, will be characterized by the massive use of air forces, various rocket weapons, and various means of mass destruction, such as atomic, thermonuclear, chemical and bacteriological weapons." Security advisors took this statement as evidence of Soviet biological weapons capacity and its intention to use them for mass destruction in future wars.

By 1956, a revised biological and chemical weapons policy was formulated to the effect that the United States would be prepared to use such weapons in a general war to enhance military effectiveness. The decision to use them would be reserved to the president. That year, the US Army manual The *Law of Land Warfare* deleted reference to "retaliation only" and asserted in stead, "The United States is not a party to any treaty, now in force, that prohibits or restricts the use in warfare of toxic or nontoxic gases, or smoke or incendiary materials or of bacteriological warfare." In the US Army manual that articulated doctrine for all the armed forces, reliance on presidential authority was clearly expressed, leaving latitude for offensive development as in the nuclear program: "The decision for U.S. forces to use chemical and biological weapons rests with the President of the United States:"

A major effect of this policy change was the expansion of plans for what biological weapons might accomplish. Large-scale simulations took on more importance. At Dugway, $2.8 million was invested in an improved grid system and meteorological network to accommodate large-scale aerial bombing and spray tests. In 1955, at the tripartite meetings with the United Kingdom and Canada, the US government announced its new approach, the "*Large Area Concept*" (LAC), whereby an airplane or a ship could spread an off-target line of pathogens and possibly infect an entire region. One major supposition was that meteorological conditions could be predicted for attack purposes, and the other was that pathogens could survive for long periods while airborne.

In 1957, to test the first supposition, UK scientists embarked on field trials in which fluorescent zinc cadmium sulfide particles were released from an aircraft flying upwind of a target area. The first tests involved releasing particles along a three-hundred-mile line over the North Sea and then taking air samples in England and Wales to track their dispersal. It was found that millions of people could be exposed to a biological aerosol. In 1957, as part of Operation LAC, the US military also disseminated particles of zinc cadmium sulfide along a path from South Dakota to Minnesota. A repeat trial in February 1958 at Dugway Proving Ground showed that six-hundred-mile dispersal could affect millions. Further tests with live simulants suggested that some pathogens could, in fact, survive these tests.

## Expansion in the Kennedy-Johnson Years

The extent to which any head of state is closely advised of biological weapons policy concerns historians of every program. The US postwar chemical and biological weapons programs were small enough, within the context of the Defense Department, to escape attention from high civilian officials. In a 1960 press conference, President Eisenhower was asked whether a policy change toward first use had occurred. From the military perspective, the change had taken place. Yet Eisenhower replied that no such official suggestion had been made to him, and he added, "so far as my own instinct is concerned it is to not start such a thing as that first."

Shortly after President John F. Kennedy's 1961 inauguration, Secretary of Defense Robert McNamara began organizational changes to centralize and streamline Pentagon authority. One result was that the biological weapons program achieved more autonomy from the Chemical Corps. It was placed under the newly formed Munitions Command, within Army Materiel Command. The renamed Fort Detrick, its center, maintained operational control over the Pine Bluff plant. Only the Army had the right to work on highly infectious agents, while the Air Force and Navy kept their research and procurement programs, estimated to account for percent of overall chemical and biological spending, which in the next several years reached an annual $300 million. Defense also instituted a Joint Technical Coordinating Committee to review relevant programs and coordinate research, but none of this activity constituted a high-level review.

The 1960s accelerated the pace of biological-weapons field testing. The escalation of the Vietnam War increased military authority over weapons development, and the CWS benefited from this expanded latitude. This period saw a great increase in the variety and scale of simulated civilian targeting, reinforced by improved information on dose response from human subjects research and by other research and technical improvements that came from numerous university and privately contracted projects. From November 1962 to April 1967, for example, Dugway Proving Grounds was nearly constantly involved in large-scale field tests of *P. tularensis*. Some tests had defensive purposes as well. In 1965, BG spores were dispersed at National Airport and Greyhound Terminal in Washington, DC; in the New York City subway in 1966; and off the coast of California, near San Clemente, where President Richard Nixon had his West Coast retreat, in August through September 1968. In 1964-65, simulated "*anti-animal*" attacks were staged at six major stockyards in Texas, Missouri, Minnesota, Iowa, Nebraska, and South Dakota. Most tests, however, were to demonstrate regional attack scenarios. Between June 1961 and November 1969, twenty-two fluorescent particle trials were staged in public domain areas, such as national forests in South Carolina and Minnesota, and in Fort Wayne, Indiana; Corpus Christi, Texas; and, again in 1968, San Francisco. St. Louis was retested three more times, from May to September 1963, April to October 1964, and March 1965, in a recapitulation of the seasonal experiments of the St Jo project.

An important innovation within the biological program was Project 112, a land and sea project for expanded testing of chemical and biological weapons. In 1962, the Army established the new Deseret Test Center, at Fort Douglas, Utah, from which the project was coordinated. The Army, Navy, and Air Force were involved; Canada and the United Kingdom were partners. Although two tests in 1968 simulated anthrax attacks on warships and how the crews might be protected, the majority of Project 112 trials appears to have been for testing the offensive capacity of biological weapons.

At least fifty Project 112 trials took place, most of them from December 1962 to 1970. Public information remains incomplete on these and the related sea trials for which the backdrop was the escalating Vietnam War. Eighteen of these trials involved simulants of biological weapons (usually BG), nearly all in the 1963-65 period. The years 1965 to 1967 were peak ones for chemical weapons tests, with a total of fourteen trials of VX, sarin, nerve gas stimulants, and tear gases. In 1967 and 1968, tests of chemical agents were staged at Porton Down and at Ralston, Canada. Otherwise, biological and chemical weapons tests were often conducted at the same sites or regions, at different times.

The sea trials conducted under Project 112 were collectively called *Shipboard Hazard* and *Defense* (SHAD). SHAD involved at least thirteen war ships, along with bombers and airplanes fitted with spray generators. One sea trial, named "*Big Tom*' was the epitome of what the Army intended with Project 112. Held off the coast of Oahu, Hawaii in May-June 1965, it brought together the Navy and Air Force to "evaluate the feasibility of a biological at tack against an island complex and to evaluate doctrine and tactics for delivery of such an attack." The test consisted of a series of off-target releases of both liquid BG (from a spray tank mounted on a Navy A-4 aircraft) and dry BG from a spray tank mounted on a USAF F-105 aircraft. The experimenters were concerned with area coverage and in particular with how much aerosol would penetrate the jungle canopy. In 1964, another trial, called *Shady Grove*, was conducted near Johnston Atoll, eight hundred miles west of Hawaii, where the US military also stored chemical weapons. Shady Grove tested actual pathogens, for *tularemia* and *Q fever*, on rhesus monkeys in cages on the decks of ships.

The development of offensive strategies for the Vietnam War was a likely goal for some of the tests. For instance, the trial "*Yellow Leaf*" was to test competing Army and Navy biological weapons systems against targets in a jungle environment. The trial began in February 1964 with a series of 184 bomb tests at the Fort Sherman Military Reservation in Panama. Political upheaval required moving the tests to the Olaa Forest in Hawaii, where an other hundred tests were staged. Also at the Olaa Forest, the Army conducted twenty aerosol tests to gauge diffusion beneath a jungle canopy. In 1965, an other trial, called "*Magic Sword*" was conducted to see if mosquitoes released in a Pacific Island setting might convey infectious diseases such as dengue fever or yellow fever.

Other field tests took place near Fort Greely, Alaska, to gauge biological warfare under USSR winter conditions. One test called "*Red Cloud*" was con ducted from late November 1966 to mid-February

1967, to evaluate the dissemination of wet and dry forms of the agent of *tularemia* in extremely frigid conditions. The tests expanded to the Pacific, with a variety of simulants and pathogens, to show that the program was ready for war.

In September and October of 1968, the final and probably most elaborate trial (DTC Test 68—50) took place at the Eniwetok Atoll in the Marshall Islands. *Staphylococcal enterotoxin*, Type B (SEB, an incapacitating toxin produced by a common bacterium, *Staphylococcus aureus*) and BG were dispersed by F4 Phantom jets with newly designed dissemination tanks. The conclusion from this trial was that a single tank could disperse a virulent infectious agent over nearly a thousand square miles.

With every test, the US military increased its readiness for strategic strikes with biological agents. During the 1960s, unfettered by any retaliation-only policy, the Army showed that it had met the challenge of large-scale attack, at least by its own calculations. It appeared only to await a command from a higher authority. That command never came. In 1969, the program's enormous effort to show that biological warfare could rival nuclear warfare—the laboratory and human subjects research, the industrial production and stockpiling of agents, the manufacture of bombs and spray generators, fitting of airplanes and ships for dispersal, the indoctrination of troops, and the large-scale field trials—would end abruptly with a brief presidential statement about the danger these weapons posed to humanity.

# 3

# Threat Created by Bioweapons

To overcome the threat of the intentional use of germs against us, we must call upon all the resources available to us. As citizens, we have the responsibility to insist that our government, both local and federal is responding appropriately to the new challenges of *bioterrorism*. Knowledge is power as we deal with *bioterrorism*. But true success requires a government that is also prepared.

As a nation, we need a cohesive, comprehensive, three-pronged approach to prevent, prepare for, and respond to any future bioterrorist attacks. Fortunately, we are not starting from scratch. A lot of work has been done in recent years to recognize and begin to deal with the threat we face. The original legislation, enacted almost a year before the September 11 attacks, provided the framework to coordinate the response of federal, state, and local agencies to a bioterrorist attack. It also provided a blueprint for strengthening the capabilities of those entities and organizations in your community—and in communities all across the country—that will be called upon to respond immediately if a bioterrorist attack occurs. These include public health agencies and laboratories, hospitals and other health care facilities, and the emergency personnel you will call and who will be responding to your calls.

But as often happens, the legislation was never fully funded. Our hope was to gradually and carefully implement the program so that over a decade, each state—and through each state, every community—would become fully prepared.

Now we know that we cannot wait. Our time line must be accelerated for the safety and health of your family. We must act now to ensure that our public health system is able to respond to what you and your family will need in the event a bioterrorist assault occurs in your neighbourhood. And that means we must provide adequate resources to communities around the country so that they can use them in their own way to meet their own particular needs, rather than providing a one-size-fits-all solution from Washington.

For these reasons, the Bioterrorism Preparedness Act of 2001, which builds on the foundation we laid two years ago with the Public Health Threats and Emergencies Act and points the way to a safer, more secure future. More than two dozen of my Senate colleagues made major contributions to this legislation. Sens. Judd Gregg, Susan Collins, Pat Roberts, Chris Dodd, and Mike DeWine deserve special recognition The legislation not only fills the gaps in our public infrastructure that were defined so clearly in the anthrax attacks in the fall of 2001 but also addresses the areas of food safety and agroterror that have been sorely neglected in the past.

## Prevention Bioterrorist Attacks

There are several things we need to do. Improving our intelligence-gathering capabilities so that we can thwart planned attacks before they happen obviously is extremely important. And our government is doing that. Who has the smallpox virus? Which rogue nations have active offensive-bioweapons programs under way?

Our war against the Taliban in Afghanistan was not only to bring to justice those responsible for the unspeakable attacks of September 11. Part of what we sought to accomplish was to wipe out their ability to stage future attacks, including bioterrorist attacks.

We know that Iraq had a substantial bio-weapons program in the years leading up to the Gulf War. Working with the United Nations, we are closely monitoring what happens there to ensure that that technology is never used.

Making sure the expertise and technology developed in the former Soviet Union's massive offensive *bioweapons* program is not made available to terrorists or rogue nations is another preventive step we are currently taking Where are the Russian scientists now?

Here at home, we need to gain tighter control over those biological agents that could be used to harm our people. We also need to greatly enhance our food safety program by giving the Food and Drug

Administration (FDA) the additional resources and authority it needs to do its job.

## THREAT CREATED BY THE BIOWEAPONS

Currently, anywhere from eleven to seventeen countries have *biological weapons programs*, including Russia, which has the largest and most sophisticated facilities, many of whose locations are unknown.

We know that, beginning in 1980, the Soviets embarked on a successful offensive bio-weapons program that produced tons of *smallpox* virus for use in bombs and intercontinental ballistic missiles. We know that the Soviets also manufactured *plague* for use in weapons and researched other biological agents, including all those discussed already in this book, such as *anthrax*, *tularemia*, and *botulinum toxin*.

With the breakup of the Soviet Union, there remains a real concern that all of that expertise and technology could fall into the hands of rogue nations or terrorist organizations. More than seven thousand scientists worked in the bioweapons program during the 1980s. Many of them are now unemployed or working in less lucrative or challenging fields, and there is the very real concern that they could offer to sell their services and their technological know- how to the highest bidder.

So under the Department of Defense's Former Soviet Union Threat Reduction program, the federal government is tairing steps to ensure that that does not happen by assisting the republics of the former Soviet Union, and in part by directly subsidizing these scientists. In 2001, we spent about $511 million to:

1. Provide contracts or grants to aid the elimination and the safe and secure transportation and storage of biological, chemical, nuclear, and other weapons.
2. Establish programs to prevent the proliferation of weapons, weapons components, and weapon-related technology and expertise.
3. Support programs that train former Soviet defense and military personnel for non-military work, and protect weapons, weapons components, weapons technology, and expertise.

It is in our country's best interests to keep Soviet scientists and technology off the black market. This program is designed to do just that the special relationship that exists between President Bush and Russian president Vladimir V. Putin will lead to new opportunities for greater cooperation and more collaboration to reduce common threats.

Many industry and law enforcement personnel have been surprised to discover how widespread is the distribution of bacteria and germs that could be used as bioweapons. Anthrax exists in laboratories of research academic, and military institutions all over the country.

Federal law already regulates the transfer of dangerous biological agents. But we need to do more. The secretary of health and human services should come up with a list of biological agents and toxins that pose the greatest threat to public health, and the government should more aggressively regulate possession, use, and transfer of those agents. Violators should face stiff civil penalties, as well as criminal sanctions.

## Protection of Food from Bioterrorists

We absolutely must do more to ensure the safety of our food. Our ports are particularly vulnerable. We currently have 175 FDA inspectors to cover more than three hundred ports where food is brought into this country. Clearly, we need to hire more inspectors, and we need to grant them the authority they need to do their job right.

In addition to hiring more inspectors, we should allow the FDA to use qualified employees from other agencies to help conduct food inspections. And any domestic or foreign plant that makes or processes food for use in the United States should have to register with the FDA so we will know where the food is coming from.

In addition, importers should be required to notify the FDA about their country of origin and what food they are bringing in. And to prevent "*port shopping*," the FDA needs to be able to mark food shipments denied entry at one port so the ships don't just sail a few hundred miles to gain entry at another port.

To prevent possible contamination of our food, the FDA should be given what is called detention authority. Instead of having to wait for people to die or become seriously ill to issue a food recall, the FDA should be able to detain a food shipment it believes poses a serious health threat to people or animals. Detention authority would ensure that the food is not distributed to your dinner table while the FDA is spending time going to court to obtain a seizure order.

We also need to give the FDA the tools it needs to trace the source and distribution of any food that poses a health threat. We can do that by simply requiring those who manufacture, process, pack, transport, distribute, receive, hold, or import food to maintain proper records for the FDA's inspection. It may be more record keeping, but it will mean that we can quickly identify any risk to your health. And anyone who engages in a pattern of trying to import food that presents a health threat to people or animals should be barred from doing business here. Period. As you can see, we have not done as good a job as we should have in giving the FDA the tools it needs so our families can rest assured that our food supply will be safe. Now is the time to act.

## Are We Prepared for a Bioterrorist Attack?

Clearly, even America is not ready to meet this deadly threat. We can send an e-mail message across the globe in the blink of an eye, but the public health agency in your community upon which you will depend might not even have equipment as basic as a fax machine. And the success of public health agencies in fighting infectious disease or bioterrorism depends on rapid communication. Scientists have deciphered the entire DNA sequence of the human genetic code, but many public health laboratories cannot conduct basic tests to identify deadly microbes rapidly and accurately.

In a disease emergency, swift action is essential to keep a local outbreak from becoming a national epidemic. Lost hours mean lost lives. Every moment counts.

Today we remain highly vulnerable. But I want to make one thing clear. It's not that we are unprepared. It's that we are under-prepared. We can't lose sight of the fact that in our federal, state, local, and private health systems, we have all the pieces we need to wage an effective defense against bioterrorism.

We just need to coordinate those pieces in a seamless way. And we need to commit the resources necessary to hire well-trained epidemiologists, upgrade our laboratory capabilities, and improve our communications systems. This can be done. But it will take your support. You must call for it.

The gaps we now see in our public health system are the result of twenty years of benign neglect and underinvestment. We have already begun to correct the problems. But we still have a lot of work to do. And we must do it now. Success will depend on the support of individual Americans who take time to understand where the gaps are and then actively call for filling the gaps.

## Bioterrorist Attack

First, investing in upgrading our public health system makes sense even if we never experience another bioterrorist attack. And I pray we won't. In a sense, the fundamental health issue here is infectious disease. Bioterrorism, when you come right down to it, is simply the deliberate attempt to spread infectious diseases. But we are going to have to deal with these and similar diseases whether they are spread by terrorists or occur naturally. This is an important point: Any improvements we make in our ability to respond to a bioterrorist attack will serve the very valuable dual purpose of better preparing us to handle all infectious and communicable diseases—from the flu that strikes millions and kills thousands each year, to the food poisoning that annually hospitalizes over 300,000, to the new emerging infectious diseases caused by antibiotic-resistant organisms, to HIV/AIDS, which infects more than a million Americans right now.

Infectious diseases are diseases caused by living organisms, such as bacteria or viruses. As we have seen, the biological agents we face are not new. Anthrax, smallpox, the plague—they have been around for centuries. And as we have also seen, there is nothing new about mankind's attempts to turn these infectious diseases against others. We are just dealing with a higher level of sophistication and technology in the way it's done.

In the last century, modem medicine made significant strides toward freeing us from the threat of many infectious diseases that had plagued humanity through the ages. We actually eradicated natural outbreaks of smallpox. That's an amazing accomplishment, a remarkable public health victory that has saved the lives of millions of children, women, and men.

Through the development of vaccines and antibiotics over the past sixty years, it appeared we were on the verge of winning the war against infectious disease. In the 1970s even the surgeon general of the United States said that we would see the end of infectious diseases as a significant problem. But in underdeveloped nations around the world, horrible infectious diseases such as Ebola continued to claim lives. And in developed countries, as antibiotic usage grew, new antibiotic-resistant organisms began to emerge.

As we move into the twenty-first century, the reality is that infectious diseases are still the second leading cause of death worldwide. Worldwide, HIV/AIDS claims three million lives a year, tuberculosis two million, and malaria one million. And here in the United States, infections remain the third leading cause of death.

Meanwhile the world, because of advances in transportation, is a much smaller place. And germs know no borders. Early last year; Bono. U2's lead singer has been very active in the global fight against AIDS and other infectious diseases. As he said in that meeting, "An infectious disease is just one cheap airplane seat away from any major city in the world." Sadly, the same can be said of bioterrorism,

## Key Areas in Public Health

As we learned from the anthrax-related events last fall, we must improve training, communication, and coordination between the various parts of our public health system.

Public health offices at the local and state levels have been under funded and understaffed for years, in part because they have not had a very effective constituency arguing for their importance in Washington, D.C. The quickest and most effective way to fill the gaps uncovered by the anthrax threats is to immediately expand federal grants to allow states and local public health offices to address the challenges that face them.

## Training

If a bioterrorist attack occurs, those on the front lines won't only be firefighters and police officers. They will be doctors and nurses in emergency rooms and private practices, and epidemiologists and lab-technicians in public and private laboratories.Their ability to quickly and correctly recognize diseases with which they have had little or, more commonly, no direct experience—anthrax smallpox plague, and the other agents we have discussed—will largely determine how many people fall seriously ill or die.

We need to better educate medical professionals about infections and organisms that could be used as bioweapons. Our intelligence community has identified which ones those are. And a better under standing of, and more research into, infectious diseases generally would better prepare us to reduce their toll of death and sickness.

## Communications and Coordination

We need to improve communication among the various clinical facilities, laboratories, and personnel in your community. We must also strengthen communication between the local public health system and the state and federal governmental bodies. The flow of information needs to be seamless.

Just after September 11, for example, the Centers for Disease Control and Prevention (CDC) sent an alert on bioterrorism to all state health officials. A week later, the alert still had not reached many local emergency rooms—which, in all likelihood, would be the first line of defense if such an attack occurred. That's inexcusable in this day of on-line access, especially when it comes to our safety.

If we are at war, the soldiers on the front line need to be the first to know that an attack is coming—not the last. And our hospitals must plan for how they would meet such a threat. Before September 11, for example, only one in five U.S. hospitals had bioterrorism - preparedness plans in place.

One of the most effective things we can do is to ensure that we develop and maintain an integrated, state-of-the-art, real-time computer communications system that efficiently links local, state, and federal public health officials and private facilities, researchers, and providers and allows them to share information with each other.

Communications between doctors and public health agencies, and between public health agencies and the government, also are lacking. One in five public health offices does not have e-mail. In this high-tech age, many public health offices still collect information on disease outbreaks by having doctors send in postcards.

Disease *surveillance* is a crucial part of the system. If ten people go to ten different doctors or hospitals with symptoms of a particular microorganism, we need to be able to make that connection faster than by sending a postcard through the mail. As we have seen time and time again in this book, every moment counts, and a fast response is literally a matter of life and death when it comes to a bioterrorist attack. The first line of defense, is alert, well-trained doctors and nurses. But this is an area where computers can help tremendously. There are data-collection systems that can analyze medical records and, in a heartbeat, make connections that people would probably miss. New York City, for instance, keeps hourly tabs on the sale of antidiarrhea medication. The city also tracks 911 calls for reports of' respiratory problems, which has proven to be a fast and accurate indicator of flu outbreaks.

We also need to improve coordination between the agencies and departments—federal, state, and local—that will respond if there is a bioterrorist attack. This should be accomplished by designating an assistant secretaiy for emergency preparedness at the Department of Health and Human Services who would be responsible for coordinating all emergency preparedness efforts including response to bioterrorist threats and attacks. And we need to bring together the leaders of key federal departments on an ongoing basis, including Health and Human Services, Justice, Defense, Agriculture, and the Federal Emergency Management Agency, to coordinate bioterrorism countermeasures, from research on likely biological agents to response to an actual attack.

Our federal departments should not duplicate each other's efforts. They should complement and supplement each other.

As we all witnessed during the *anthrax* scare last fall, we must improve and streamline our methods of communicating with the public. As part of this strate the federal government and the states should designate in advance a lead spokesperson who is able to reliably and accurately convey vital information to the public in the event of an attack. We also should provide uniform, up-to-date information through an official federal website and explore other means of rapidly communicating accurate information to the public.

The bioterrorist events of last fall underscored the critical need for—and the value of—accurate and timely information. It leads to a more responsive citizenry and a calmer nation. When we don't know, we should say, "We just don't know yet."

## Special Focus

As we develop plans to deal with the threat of *bioterrorism*, we must specifically focus on the needs of children and other vulnerable populations. As we have seen, children may be at particular risk during a biological or chemical attack. Many of the vaccines and antibiotics we would use have not been studied adequately in children. While there clearly is much more we need to learn about the impact of various biological agents, our knowledge about how they affect children and how to treat children lags even further behind.

Moreover, as suggested by the anthrax attacks of last fall, the immune system of elderly individuals may be sufficiently different to increase their susceptibility to certain microorganisms.

Much of what needs to be done is outlined above in terms of better planning and enhanced capabilities. But states also need to reexamine their basic public health laws to make sure they are updated to deal with current threats. At the request of the CDC, the Center for Law and the Public's Health in 2001 drafted a model piece of legislation called the State Emergency Health Powers Act. This model law identifies areas on which states should focus to prepare for the threat of bioterrorism, as well as other public health threats and emergencies. It calls for states to create a public health emergency planing commission charged with developing a comprehensive plan to cover all aspects of a public health emergency, including how to:

1. Notify and communicate with the public during an emergency
2. Centralize coordination between state, federal, and local agencies

## FROM A DARK WINTER TO A BRIGHT FUTURE

In the past two years, we have held two military exercises to test our readiness to respond to a bioterrorist attack. We must heed the lessons learned from those exercises.

TOPOFF, short for "*top officials*," was held in March 2000 and simulated a plague outbreak in Denver. Within a week, the hypothetical plague had spread to other states, and even as far away as England and Japan. By the end of the exercise, the number of sick and dead could not be determined because of conflicting reports, but the range was 3,700 to more than 4,000 cases, with 950 to more than 2,000 deaths.

Among the key results that we can—and must—build on were:

1. It was unclear who was in charge. Conference calls, some of which included fifty to a hundred people from various federal, state, and local agencies who had never previously met or worked together, bogged down in indecision.
2. Communications broke down. Information on such crucial issues as how many confirmed or suspected cases of plague had broken out would not have been available to decision makers. Hospital officials could not reach health department and other officials because they were not answering their phones: they were at operations centers, or taking part in conference calls, or were just too busy managing the crisis.
3. Delivery and distribution of antibiotics to hospitals was delayed. Local supplies were exhausted early in the crisis. When antibiotics were sent from the National Pharmaceutical Stockpile, they sat at the airport for hours while plastic bags were bought and delivered, and then one person counted out the pills and put them into the bags.
4. Health care facility as were swamped. Within the first twenty-four hours of the exercise, hospitals had exceeded their capacity. And much of the demand came from "worried well"—people who didn't have the plague but were afraid that they might.

Dark Winter was held at Andrews Air force Base in June 2001. It simulated a smallpox outbreak. Within two weeks, the hypothetical epidemic spread to twenty-five states and ten other countries, with 16,000 cases and 1,000 deaths.

Among the key findings were:

1. Leaders were unprepared for how a bioterror attack would unfold. Senior decision makers, particularly in defense and national

security, were unfamiliar with public health strategies and options to deal with the crisis.

2. Crucial information was lacking. Once again, such vital information as how many people had been exposed, how many were hospitalized and where, and how many had been vaccinated was unavailable to decision makers.
3. Options were limited by the lack of sufficient vaccine supplies. Shortages led to violence and flight by those desperate to get vaccinated.
4. The health care system lacks the surge capacity needed to deal with such a crisis. Hospitals were quickly overburdened. As with TOPOFF one of the greatest challenges was dealing with the huge numbers of worried well—those who showed up at hospitals (or treatment even though they did not have smallpox, This made it far more difficult to treat those who actually were ill.
5. It was unclear who was in charge. Conflicts broke out between state and federal leaders over who would decide such crucial issues as whether there should be mandatory isolation end vaccination.
6. Leaders must gain the public's trust and cooperation to contain disease. It's imperative that decision makers clearly convey to the public that distribution of vaccine and other disease containment efforts were fair and based on the general good of society.

As we look to the future, we must fill the gaps identified by these two instructive exercises.

1. Provide essential emergency needs, including medical supplies, drugs, vaccines, food, shelter, and beds
2. Keep courts operating smoothly, and even provide for emergency judges in case quarantining is required
3. Safely evacuate, feed, and house people in an emergency, if needed
4. Train health care providers to diagnose and treat infectious diseases
5. Develop guidelines for vaccinations

Once the plan is in place, it would grant the governor of a state broad powers to declare a public health emergency when there is an imminent threat or outbreak of infectious disease, whether caused by bioterrorism or occurring naturally.

The governor would have the authority to use all. of the resources of state government to respond, including transferring personnel from other state departments and agencies to help where needed. The National Guard also could be called to active duty.

In the military modeling exercises TOPOFF and Dark Winter; we saw the problems we could face with inadequate personnel and supplies. We also saw the potential for civil unrest if the public is not clear about who's in charge and why actions are being taken.

The model legislation spells these things out clearly. Rationing of vaccines or other supplies, if necessary, would be controlled by public health officials. Preference could expressly be given to health care providers, disaster response teams, and mortuary staff. Public health officials would have broader powers to compel individuals to undergo examination, medical testing, vaccination, or treatment.

As part of those broader powers, officials would have more options to quarantine those who are determined to pose a serious threat to public health. Several steps would be required to make that determination, as officials must balance the individual's rights against dangers to public health. And protections must be built in to ensure patient confidentiality.

Enactment of this general type of legislation would go a long way toward ensuring that we are adequately prepared for the threat of bioterrorism. But governors and other officials need to use forbearance and caution in exercising some of the sweeping powers that would be provided by such laws. And remember that legal authorities alone are no substitute for ongoing training and adequate public health resources. I also want to stress that this legislation is needed to deal with all health risks posed by infectious diseases.

## Response to a Bioterrorist Attack

As the military exercises clearly demonstrated, our health care system is not capable of handling what's called the surge that would be created by a large-scale biological disaster. In some hospitals now, a flu outbreak can overtax resources. This group of volunteers could be called up by the secretary of health and human services to respond to a public health emergency, much like the National Guard is mobilized for disasters. We also need to increase resources for the National Pharmaceutical Stockpile so that we are fully prepared for any attack. This valuable national resource is ready today to ship at a moment's notice truckloads of lifesaving vaccines and medical supplies and equipment so that they arrive within twelve hours anywhere in the United States. The pharmaceutical stockpile was used for the first time in response to the September 11 attacks in New York City.

There are still infectious diseases for which we don't have vaccines or treatments, so we need to devote more resources to developing them. Because early testing on human-beings is obviously out of the question, the FDA needs to come up with reasonable guidelines for the approval of potentially lifesaving drugs based on animal testing. And to make sure we have sufficient supplies of vaccines. We favour having the federal government enter into long-term contracts with suppliers guaranteeing the purchase of a certain quantity at a fixed price. To encourage participation, the private sector needs some certainty in terms of market. New public- private partnerships are the only way to speedily address the development of new tests, vaccines, and treatments to combat the increasing threat of bioterrorism.

We also should provide limited antitrust relief to drug companies so they can use resources wisely to quickly produce needed vaccines and drugs. And we need to extend reasonable limits on liability exposure so that these countermeasures can be produced at a reasonable cost. Reasonable liability protection is an important component of the public-private partnership and is the only way that our country can deliver in a timely way the vaccines that will protect you and your family.

## Challenges

Absolutely! We know the challenges. They were not completely unknown before the anthrax attacks of 2001, but they were not nearly so clearly defined. We now call upon the best in the American spirit that exists in communities all across this great land. As Americans, we have always looked to the future. With knowledge of what to do, we know that we can be prepared. With an understanding of what to do in the event of an attack, we will be protected. With recognition that we have underinvested in our public health infrastructure, we will call for improvement.

And with the most powerful private and public research and scientific communities in the world, the synergies of partnerships in addressing the threats before us will lead to success. Partnerships between federal and local governments, partnerships among the federal agencies, and partnerships among local public health entities are being established to the benefit of all. In the coming months, we will continue to assess where we are and what we need to do to be fully prepared. Our benchmark will be the changes that we incorporate to ensure that each day, we are more able to respond because we have thought about our vulnerabilities, assessed appropriate actions, and taken steps to ensure that the next biological attack will be met with the full force of a coordinated, well-developed, expertly trained disaster response team. Every one of us is a key player on that team.

# 4

# AIRBORNE DISEASES

The produce looked good that day. The lettuce was green and crisp, and the carrots just seemed to exude good health. The broccoli was a deep, forest green and the cauliflower a lily white. It was probably due to the ultrasonic humidifier spraying its mist over the vegetable section. That fresh, clean mist made everything look good.

So it was that November day in Bogaloosa, Louisiana, when officials from the local Department of Health walked into the grocery store. They had a disturbing story: 28 people were sick with Legionnaires' disease and 2 had already died to disease-related pneumonia. Most of the patients, it seemed, had been in the grocery store within the previous ten days.

The investigators were friendly but firm: how long, they asked, did the air conditioner run each day? Were the vegetables washed before putting them out and if so, for how long and where? Had any of the employees been sick recently? Who was the grocery store's supplier? And that humidifier—how long had it been there?

Three days later the investigators were back with the answers. It was the humidifier! The lab had found *Legionella pneumophila* in the water in its reservoir. Could the owner please remove it from the store and clean it out with disinfectant before reusing it? Within a day the humidifier was gone. The broccoli did not look quite so good, and the lettuce leaves seemed to droop. But the air was after to breathe, and that was important.

Though known only since 1976, Legionnaire's disease is recognized today throughout the world, and outbreaks such as this 1989 incident in Louisiana are reported regularly to the CDC. In the United States alone, microbiologists estimate that between 25,000 and 50,000 cases occur annually, though most go unreported. Legionnaires' disease is one of the contemporary airborne bacterial diseases that we shall discuss in this chapter. Such diseases are generally spread where people crowd together. In many instances, reduced body resistance is a prime motivating factor in the establishment of infection.

We hall survey the airborne bacterial diseases according to two general categories. The first category will include diseases of the upper respiratory tract such as strep throat, scarlet fever, diphtheria, pertussis, and several forms of meningitis. The second category will include diseases of the lower respiratory tract: tuberculosis, penumococcal pneumonia, primary atypical pneumonia, Legionnaires' disease, and others. As we proceed, note how the initial focus of infection may be a point from which the bacteria spread to other organs. Also note that antibiotics are available for the treatment of these diseases and that immunizations are used to protect the community at large.

## Diseases of the Upper Respiratory Tract

### Streptococcal Diseases

Streptococci are a large and diverse group of encapsulated Gram-positive bacteria. They cause streptococcal sore throat (strep throat), scarlet fever, and several other diseases in humans. The bacteria divide in one plane and cling together to form chains of various lengths. This observation is apparent in streptococci grown in liquid media, but it may not be obvious in streptococci from solid media.

Microbiologists classify the numerous types of streptococci by several systems of which two are widely accepted. The first system, developed by J.H. Brown in 1919, divides streptococci into hemolytic ("blood-digesting") groups, depending on how they affect red blood cells. For example, *alpha-hemolytic streptococci* turn blood agar an olive-green colour as they partially destroy red blood cells in the medium. By contrast, colonies of *beta-hemolytic streptococci* are surrounded by clear, colourless zones due to the complete destruction of red blood cells; and *gamma-hemolytic streptococci* have no effect on red blood cells and thus cause no change in blood agar.

The second classification system, suggested by Rebecca Lancefield in 1935, is based on variants of a specific carbohydrate, the C substance, in cell walls of streptococci. Group A through O streptococci have been distinguished. Most streptococcal diseases in humans are caused by species of Group A streptococci; *Streptococcus pyogenes* is the most common species. This beta-hemolytic organism is the one generally meant when physicians refer to "Group A beta-hemolytic streptococci."

*Streptococcus pyogenes* is the cause of streptococcal sore throat, popularly known as *strep throat*. *S. pyogenes* is one of the few bacteria that can invade the normal tissues of the pharynx and cause a *primary disease*, that is, one developing in an otherwise healthy body. The streptococci reach the upper respiratory tract within airborne *droplets*, the tiny particles of mucus expelled during coughing and sneezing. After a short incubation period (one to three days), a strep throat develops. Patients experience a high fever, coughing, swollen lymph nodes and tonsils, and a fiery red "beefy" appearance to pharyngeal tissues owing to tissue erosion. About 500,000 Americans suffer strep throat annually, many cases complicated by infection in the middle ear, *otitis media*.

This pathogenicity of *Streptococcus pyogenes* is enhanced by a substance called the *M protein*. This protein, located in the cell wall and pili, encourages adherence to the pharyngeal tissue and retards phagocytosis. Over 60 specific types of M protein have been identified, and complete immunity to streptococcal disease requires that a person produce antibodies against all 60 types. Since this is improbable, and individual may suffer multiple bouts of strep throat over the course of a lifetime.

*Scarlet fever* is a strep throat accompanied by a skin rash. The rash develops as a pink-red blush, especially apparent on the neck, chest, and soft-skin areas of the arms. It results from blood leaking through the walls of capillaries damaged by a toxin. This toxin, called an *erythrogenic toxin* ("red-forming"), is produced only by certain strains of *S. pyogenes*. Normally an individual experiences a single case of scarlet fever per lifetime because the immune system produces special antibodies, called *antitoxins*, which circulate in the bloodstream and neutralize the toxins in succeeding episodes.

Treatment of streptococcal diseases is generally successful with antibiotics such as penicillin and erythromycin. However, widespread resistance to antibiotics has been noted since the early 1970s, and research indicates that many strains of *S. pyogenes* carry bacteriophages having antibiotic-resistance genes.

An important complication of streptococcal disease is *rheumatic fever*. This condition is characterized by fever and inflammations of the small blood vessel. Joint pain is common, but the most significant long-range effect is permanent scarring and distortion of the heart valves, a condition called *rheumatic heart disease*. The damage arises from a reaction of the body's antibodies with streptococcal antigens bound to the heart tissue. Antibodies may also react with tissue antigens identical to streptococcal antigens. Another manifestation of this reaction may develop in the kidneys, where the condition is called *glomerulonephritis*.

*S. pyogenes* is commonly present in the human nose and throat, and transmission to many other tissues is possible. For example, streptococci may cause disease in open wounds or skin abrasions, resulting in *erysipelas*. If streptococci invade and destroy the underlying fascia over the muscles and beneath the skin, they cause a condition called *necrotizing fasciitis* and an unsightly degeneration of the skin tissues. (Tabloid newspapers and TV programs had a field day with this disease in 1994 as they heralded an outbreak of the "flesh-eating bacteria.") If the tissue of the uterus becomes infected during the birth process the mother may suffer *puerperal sepsis*, also known as *childbed fever*. Moreover, streptococci may infect the lower respiratory tract and cause streptococcal pneumonia as a complication to viral disease. In past generations, many died of *septicemia* or "blood poisoning," a disease now known to be due to several organisms, including streptococci. One of Louis Pasteur's daughters was a victim of this disease.

**Diphtheria**

*Diphtheria* was first organized as a clinical entity in 1826 by French pathologist Pierre F. Bretonneau. Bretonneau named the disease *la diphtherite* from the Greek *diphtheria* for membrane, a reference to the "membranes" that appear in the throats of patients. In 1883, Edwin Klebs observed a bacterium in material from a patient's throat, and the next year, Friederich Loeffler successfully cultivated the organism. The so-called Klebs-Loeffler bacillus has the scientific name *Corynebacterium diphtheriae* because it resembles a club ("coryne" is Greek for club) and forms membranes.

*Corynebacterium diphtheriae* is a Gram-positive bacillus containing numerous *metachromatic granules* that show up with special stains as cytoplasmic dots. The bacteria remain close to one another after multiplying and form a palisade layer, a picket-fence arrangement. The staining and reproductive characteristics are aids to identification.

Diphtheria is acquired by inhaling respiratory droplets. Infection occurs in the upper respiratory tract near the tonsils. The bacteria produce a potent *exotoxin* that interferes with protein synthesis in epithelial cells, the cubelike cells that line the skin and body cavities such as the respiratory tract. As dead tissue accumulates with mucus, white blood cells, and fibrous material, a leathery *pseudomembrane* forms ("pseudo" because it does not fit the definition of a true membrane). Respiratory blockage and death may follow, especially in children. In adults, the toxin often spreads to the bloodstream, where it causes heart damage and destruction of the fatty sheaths surrounding nerves.

Treatment of diphtheria requires both antibiotics for the bacteria and antitoxins to neutralize the toxins. At present, the number of cases of diphtheria in the United States is less than a dozen annually but the disease remains a health problem in many regions of the world, especially parts of Asia. In 1995, for example, a major outbreak occurred in 13 of the 14 new states of the former Soviet Union and affected almost 50,000 people, with about 2000 deaths.

Immunization against diphtheria may be rendered by an injection of diphtheria *toxoid* contained in the diphtheria-pertussis-tetanus (DPT) vaccine. A toxoid consists of toxin molecules treated with formaldehyde or heat to destroy their toxic qualities. The toxoid induces the immune system to produce antitoxins that circulate in the bloodstream throughout the person's life.

In 1961, V.J. Freeman discovered that a virus attached to the bacterial chromosome contains the genetic code for exotoxin production. The virus, termed a *corynephage*, exists in a lysogenic relationship with the bacterium. Strains of *C. diphtheriae* lacking the virus are harmless, but they may become pathogenic if the virus infects them.

**Pertussis (Whooping Cough)**

*Pertussis*, also known as *whooping cough*, is one of the more dangerous diseases of childhood years. Although the incidence of this disease has declined substantially from the 250,000 annual cases reported during the 1930s, the number of cases in the United States still remains significant. For example, the CDC recorded about 4000 cases in 1995, and the numbers were growing. Almost half occurred in children under six months of age.

Pertussis is caused by *Bordetella pertussis*, a small Gram-negative rod first isolated by Jules Bordet and Octave Gengou in 1906, and commonly known as the Bordet-Gengou bacillus. The bacillus is spread by droplets and uses its pili to adhere to the cilia of epithelial cells

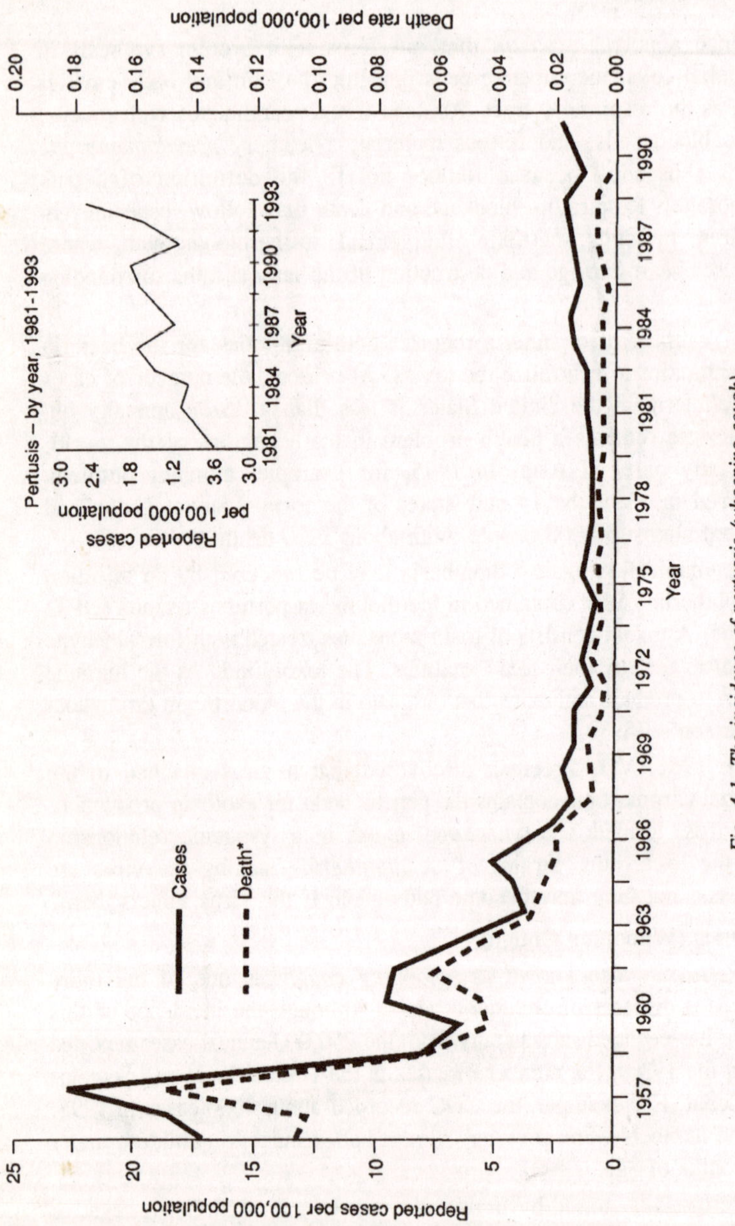


*Fig. 4.1. The incidence of pertussis (whooping cough).*

of the upper respiratory tract. There is generally no invasion of tissue but the ciliated cells are destroyed and mucus movement is impaired.

Typical cases of pertussis occur in three stages. The initial stage is marked by general malaise, low-grade fever, and increasingly severe cough. During the next stage disintegrating cells and mucus accumulate in the airways and cause labored breathing. Patients experience multiple paroxysms of rapid-fire *staccato coughs* all in one exhalation, followed by a forced inspiration over a partially closed glottis. The rapid inspiration results in the characteristics "whoop" (hence, whooping cough). Ten to fifteen paroxysms may occur daily, and exhaustion usually follows each. During the third stage, sporadic coughing may continue for several weeks, even after the bacteria have vanished.

Treatment of pertussis is generally successful when penicillin or erythromycin is administered before the respiratory passageways become blocked. Diagnosis of the disease is performed by obtaining swabs from the posterior pharyngeal wall and identifying *B. pertussis* on selective media. A fluorescent antibody test is also used. Although several toxins have been isolated from the bacillus, no one toxin is considered the prime factor in the disease.

Use of the pertussis vaccine is not without risk, however. About one in every 300,000 vaccinated individuals may suffer temporary seizures, and several may experience high fever for a short period of time. Reports such as these have raised public concern about use of the vaccine and as of 1993, public health official were recommending the newer acellular pertussis vaccine prepared from *Bordetella pertussis* extracts. Combined with diphtheria and tetanus toxoids, the triple vaccine is given the acronym DTaP. At this writing, DTaP is recommended as the fourth and fifth doses of the five-dose regimen given to children. The whole-cell pertussis vaccine in DPT continues to be suggested for the initial three doses.

## Meningococcal Meningitis

The term *meningitis* refers to several diseases of the meninges, the three membranous coverings of the brain and spinal cord. Meningitis may be caused by viruses, fungi, protozoa, or bacteria, and different forms of the disease have different mortality rates. In all forms, the meninges become inflamed, causing pressure on the spinal cord and brain. Patients experience headaches, neckaches, and lower backaches.

A particularly dangerous form of meningitis is *meningococcal meningitis*. This disease is caused by *Neisseria meningitidis*, a small Gram-negative, encapsulated diplococcus commonly called the *meningococcus*. Groups A, B, C, Y, and W-135 groups of meningococci are known. They enter the body by droplets, often from a reservoir. In most cases, the disease consists of an influenzalike upper respiratory infection. However, the infection sometimes spreads to the bloodstream, where bacterial toxins may overwhelm the body in as little as two hours and cause death. This condition is called *meningococcemia*. In survivors, the meninges become inflamed, and patients experience a characteristic stiff, arched neck and pounding headache. A rash also appears on the skin, beginning as bright red patches, which progress to blue-black spots. Fifty percent of untreated cases may be fatal.

Early diagnosis and treatment of meningococcal meningitis can prevent irreversible nerve damage or death. A principal criterion for diagnosis is the observation of Gram-negative diplococci in samples of spinal fluid obtained by a spinal tap. In active cases of disease, *N. meningitidis* may be cultivated from this fluid. Treatment with rifampin, penicillin, or sulfonamide drugs is usually recommended. A vaccine containing capsular polysaccharides from several groups has been available for immunization since 1985 but it is used only under special circumstances.

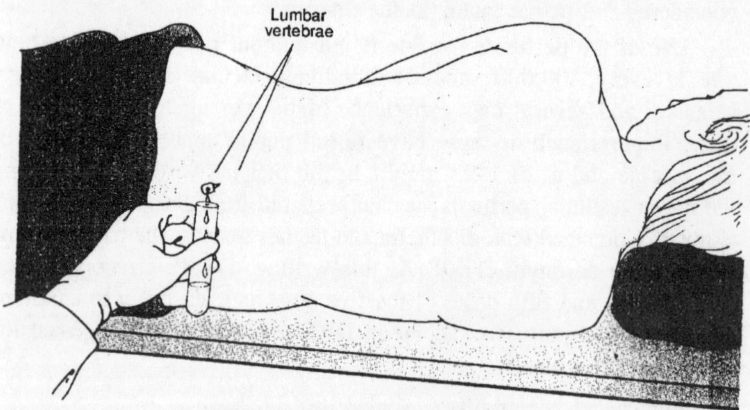


*Fig. 4.2. The spinal tap procedure.*

Cases of meningococcal meningitis are sometimes complicated by the formation of lesions in the adrenal glands. This condition, called the *Waterhouse-Friderichsen syndrome*, may be a manifestation of a hypersensitivity reaction taking place in the body. Hormone imbalances usually signal its development.

*Neisseria meningitidis* is a fragile organism that does not survive easily in the environment and must be maintained in nature by person-to-person transfer. Meningococcal meningitis is therefore prevalent where people are in close proximity for long periods of time. Grade school classrooms, military camps, and prisons are examples. Thought most people suffer nothing worse than a respiratory disease, the Centers for Disease Control and Prevention reported about 2500 cases of *Neisseria*-linked meningitis in 1994.

### *Haemophilus* Meningitis

In 1892, Richard Pfeiffer isolated a small, Gram-negative encapsulated rod he thought was the cause of influenza. Because of this relationship and due to its attraction to blood ("hemo-philus") the organism was named *Haemophilus influenzae*. However, during the great influenza epidemic of 1918-1919, Pfeiffer's bacillus was identified as a secondary cause of disease, and influenza itself was attributed to a virus rather than a bacterium.

*Haemophilus influenzae* is now regarded as an infrequent cause of respiratory tract infections, occurring primarily as a complication to a previous disease. Also, in recent years *H. influenzae* type b has attracted widespread attention as an important cause of meningitis in children between the ages of six months and two years. (The disease is sometimes called "Hib disease.") The organism moves from the respiratory tract to the blood and then to the meninges, where it causes *Haemophilus meningitis*. Symptoms of disease include stiff neck, severe headache, and other evidence of neurological involvement such as listlessness, drowsiness, and irritability. Combinations of drugs including rifampin are used for treatment because the many strains are resistant to one or more drugs. The disease has a mortality rate of about 5 percent.

In 1986, about 18,000 cases of *Haemophilus* meningitis were occurring in the United States annually and *H. influenzae* b was named the most common cause of bacterial meningitis in the United States. By that time, however, a vaccine had been licensed by the FDA. The vaccine consists of polysaccharides from the organism's capsule and is thus an acellular vaccine. As of 1993, the vaccine was combined with the DPT vaccine for distribution to children as Tetramune, and by 1994, the number of annual cases in the United States was down to 280. Indeed, the CDC had already launched an effort to eliminate *Haemophilus* meningitis by the end of 1996.

## Diseases of the Lower Respiratory Tract

### Tuberculosis

At the turn of the century, *tuberculosis* was the world's leading cause of death from all causes, accounting for one fatality in seven. Today's statistics, though improved, are still very high. In

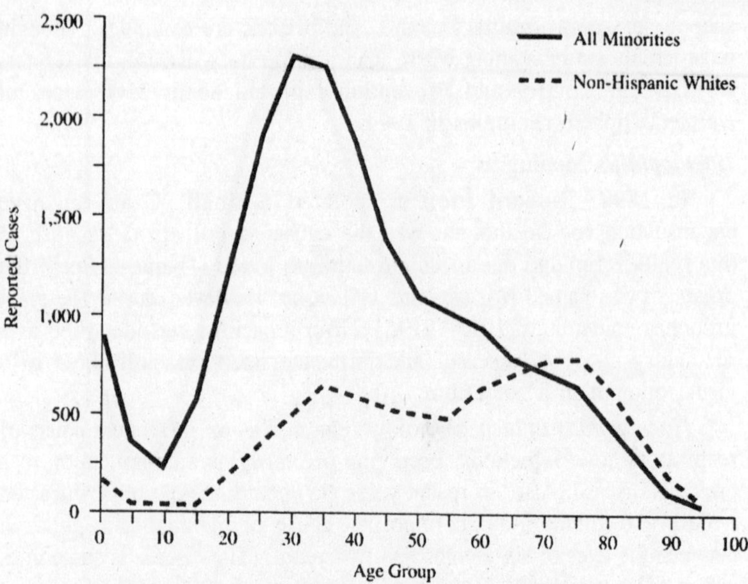


*Fig. 4.3. Distribution of tuberculosis cases by age, race, and ethinicity.*

underdeveloped countries, public health officials report more deaths from tuberculosis than from any other bacterial disease. In the United States, about 27,000 new cases are identified annually, with almost 3000 fatalities, most among individuals from minority groups. The World Health Organization has referred to the global situation as a scandal and estimates that 1.7 billion people are infected.

Tuberculosis is caused by *Mycobacterium tuberculosis*, the "tubercle" bacillus originally isolated by Robert Koch in 1882. It is a small rod that enters the respiratory tract in droplets, and multiple exposures are generally necessary. People who live in urban ghettoes often contract tuberculosis because malnutrition and a generally poor quality of life contribute to the establishment of disease, and overcrowding increases the concentration of bacilli in the air.

About 10 percent of people who contract tuberculosis become ill within three months. They experience chronic cough, chest pain, and high fever, and they expel *sputum*, the thick matter accumulating in the lower respiratory tract. Often the sputum is rust-coloured, signaling that blood has entered the lung cavity. The remaining 90 percent of victims exhibit no readily distinguishable symptoms, except malaise, fever, and weight loss. In these cases the body responds to the disease by forming a wall of white blood cells, calcium salts, and fibrous

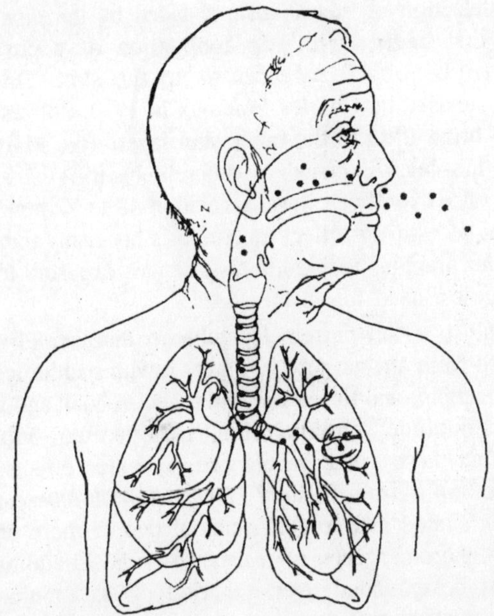

*Fig. 4.4. Mycobacterium tuberculosis is inhaled to the lungs.*

materials around the organisms. As these materials accumulate in the lung, a hard nodule called a *tubercle* arises (hence the name tuberculosis). This tubercle may be visible in the chest X ray. Despite the body response, the bacilli are not killed, and the tubercle may expand as the lung tissue progressively deteriorates.

In many instances the tubercle breaks apart and bacteria spread to other organs such as the bone, liver, kidney, and meninges. The disease is now called *miliary tuberculosis* from the Latin *milium*, for seed. Tubercle bacilli produce no discernible toxins, but growth is so unrelenting that the tissues are literally consumed, a factor that gave tuberculosis its older name, *consumption*. In children, tuberculosis of the spine is known as *Pott's disease*.

*Mycobacterium tuberculosis* contains a layer of fatty waxy material in its cell wall that greatly enhances resistance to environmental fluctuations. In the laboratory, stain must be accompanied by heat to penetrate this barrier or a lipid-dissolving material must be used. Once stained, however, the organisms resist decolourization, even when subjected to a 5 percent acid-alcohol solution. Thus the bacilli are said to be acid-resistant, or *acid-fast*. The acid-fast test is an important diagnostic screening tool when used with a sample of patient's sputum.

Early detection of tuberculosis is aided by the *tuberculin test*, a procedure that begins with the application of a purified protein derivative (PPD) of *M. tuberculosis* to the skin. One method of application, devised by Charles Mantoux in 1910, utilizes a superficial injection to bring PPD to the outer skin layers (the Mantoux test). If the patient has been exposed to tubercle bacilli, the skin becomes thick and a raised, red welt develops within 48 to 72 hours. A positive test does not necessarily reflect tuberculosis but many indicate a recent immunization, previous tuberculin test, or past exposure to the disease. It does suggest a need for further tests.

Tuberculosis is an extremely stubborn disease. Physicians once recommended fresh air but now they treat it with such drugs as isoniazid (INH) and rifampin, and to a lesser extent, ethambutol and streptomycin. Alternatives include pyrazinamide, ethionamide, viomycin, and cycloserine but these drugs have more toxic side effects. The recent appearance of *multi-drug-resistant Mycobacterium tuberculosis* (MDR-TB) has necessitated that combinations of two or more drugs be used to help delay the emergence of resistant strains. In addition, the drug therapy must be extended over a period of six to nine months or more, partly because the organism multiplies at a very slow rate (its generation time is about 18 hours). Early relief, boredom, and forgetfulness often cause the patient to stop taking the medication and the disease flares anew.

Tuberculosis is a particularly insidious problem to those who have *AIDS*. In these patients the T-lymphocytes that normally mount a response to *M. tuberculosis* are being destroyed, and the patient cannot respond to the bacterial infection. HIV-infected patients face a mortality rate from tuberculosis of between 70 and 90 percent, usually within one to four months of developing symptoms. Unlike most other tuberculosis patients, those with HIV usually develop tuberculosis in the lymph nodes, bones, liver, and numerous other organs. Ironically, AIDS patients often test negative for the tuberculin skin test because without T-lymphocytes, they cannot produce the telltale red welt that signals infection. Tuberculosis is often the first disease to occur in the AIDS patient, even before any of the opportunistic diseases appear, and it is generally more intractable than in non-AIDS patients. The WHO estimates that about 4.4 million people are coinfected with HIV and tuberculosis worldwide.

Immunization to tuberculosis may be rendered by infections of an attenuated strain of *Mycobacterium bovis*. This species causes tuberculosis in cows as well as humans. The attenuated strain is called *bacille Calmette Guerin*, or BCG, after Albert Calmette and Camille Guerin, the two French investigators who developed it in the 1920s. Though the vaccine is utilized in other parts of the world, many scientists oppose its use in the United States because they point to the success of early detection and treatment and to the vaccine's occasional side effects. Indeed, in 1989 the CDC announced a strategic plan to eliminate tuberculosis from the United States by the year 2010. The plan emphasized more effective use of existing prevention and control methods and the development of new technologies for diagnosis and treatment.

Several other species of *Mycobacterium* bear mentioning. The first, *Mycobacterium cheloni*, is an acid-fast rod frequently found in soil and water. During the 1980s microbiologists first recognized this bacillus as a cause of lung diseases, wound infections, arthritis, and skin abscesses. *Mycobacterium haemophilum* surfaced as a pathogen in 1991 when 13 cases occurred in immuno-compromised individuals in New York City hospitals. Cutaneous ulcerating lesions and respiratory symptoms were observed in the patients.

Another species is *Mycobacterium avium-intracellulare*. This acid-fast rod may be the cause of lung disease, especially in persons who have acquired immune deficiency syndrome (AIDS). In AIDS patients, the body's immune system is severely depressed, and the mycobacteria rapidly invade the lungs, often causing death. The disease is commonly called *mycobacteriosis* or *MAI disease*.

The final organism, *Mycobacterium scrofulaceum*, causes tuberculosis of the neck tissues, a disease commonly known as *scrofula*. Scrofula is manifested by growth of mycobacteria in the lymph nodes of the neck and substantial swelling of the tissues. The disease has been recognized since ancient times. Its prevalence in the Elizabethan era may have influenced people to wear the high-neck collars characteristics of that period.

**Pneumococcal Pneumonia**

The term *pneumonia* refers to microbial disease of the bronchial tubes and lungs. A wide spectrum of organisms, including many different viruses and bacteria, may cause pneumonia. Over 80 percent of bacterial cases are due to *Streptococcus pneumoniae*, a Gram-positive encapsulated chain of diplococci traditionally known as the *pneumococcus*. The disease is commonly called *pneumococcal pneumonia*.

Pneumococcal pneumonia exists in all age groups, but the mortality rate is highest among older persons and those with underlying medical conditions. *Streptococcus pneumoniae* is usually acquired by droplets or contact, and the pneumococci exist in the respiratory tract of the majority of Americans. However, the natural resistance of the body is high and disease usually will not develop until the defense are compromised. Malnutrition, smoking, viral infections, and treatment with immune-suppressing drugs typify conditions that may precede pneumonia.

Patients with pneumococcal pneumonia experience high fever, sharp chest pains, difficult breathing, and rust-coloured sputum. The colour results from blood seeping into the alveolar sacs of the lung as bacteria multiply and cause the tissues to deteriorate. Involvement of an entire lobe of the lung is called *lobar pneumonia*. If both left and right lungs are involved, then the condition is called *double pneumonia*. Scattered patches of infection in the respiratory passageways are referred to as *bronchopneumonia*. The term "walking pneumonia" has no clinical significance but is merely a colloquial expression for a long-lasting and relatively mild case of pneumonia.

The traditional drug of choice for pneumococcal pneumonia has been penicillin, with tetracycline and chloramphenicol used for people who are allergic to penicillin. However, public health microbiologists have noted an increasing incidence of resistant strains of *S. pneumoniae*, and as early as 1981 the first strain resistant to all three antibiotics was reported. In such cases erythromycin is useful.

Microbiologists have identified over 80 strains of *S. pneumoniae* based on the presence of different components in its capsule. Unfortunately, recovery from one strain lends immunity to that strain alone. In 1983, the FDA licensed a vaccine for immunization to 23 strains of the organism. Since these strains are responsible for 87 percent of cases, there is hope that pneumococcal pneumonia may be controlled in high-risk patients. In 1993, the CDC reported that an estimated 40,000 pneumonia deaths occur annually.

**Primary Atypical Pneumonia (PAP)**

During the 1940s, a significant increase in the number of unusual cases of human pneumonia prompted a search for a new infectious agent. At Harvard University, Monroe Eaton isolated a tiny, viruslike agent from the respiratory tracts of patients and cultivated it on media supplemented with blood. The organism was subsequently named *Mycoplasma pneumoniae*, the Eaton agent. Its disease came to be known as *primary atypical pneumonia* (PAP), "primary" because it occurs in previously healthy individuals (pneumococcal pneumonia often is a secondary disease), "atypical" because the organism differs from the typical pneumococcus and because symptoms are unlike those in pneumococcal disease.

*Mycoplasma pneumoniae* is recognized as one of the smallest bacteria causing human disease. Mycoplasmas measure about 0.2 μm in size and are *pleomorphic*, that is, they assume a variety of shapes. They have no cell wall and thus no Gram reaction or sensitivity to penicillin. *M. pneumoniae* is very fragile and does not survive long periods outside the human or animal host. Therefore it is maintained in nature by passage from host to host, and multiple exposures to droplets are usually necessary to establish disease.

PAP resembles viral pneumonia and other respiratory diseases in its symptoms. The patient experiences fever, fatigue, and a characteristic dry, hacking cough. Research indicates that the organisms attach to and destroy the ciliated cells lining the respiratory tract. Blood invasion does not occur, and the disease is rarely fatal. Often it is called *walking pneumonia* or *Mycoplasma pneumonia*. Epidemics are common where crowded conditions exist, such as in college dormitories, military bases, and urban ghettoes. Erythromycin and tetracycline are commonly used as treatments. Research in the 1940s established that antibodies produced against *Mycoplasma pneumoniae* agglutinate Type O human red blood cells at 4°C but not at 37°C. This observation was used to develop the *cold agglutinin screening test* (CAST) in which a patient's serum is combined with red blood cells at cold temperatures and the red cells are observed for clumping. Diagnosis is also assisted by isolation of the organism on blood agar and observation of a distinctive "fried egg" colony appearance. Experimental vaccines sprayed into the nasal passageways appear to reduce the incidence of disease, but none is licensed for use at this writing.

**_Klebsiella_ Pneumonia**

In 1882, Carl Friedlander isolated *Klebsiella pneumoniae*, a bacillus he believed was the sole caused of bacterial pneumonia. In the years thereafter, *Streptococcus pneumoniae* was found to be the major cause of pneumonia, and Friedlander's bacillus was relegated to a secondary position as the cause of about 5 percent of cases.

*Klebsiella pneumoniae* is a Gram-negative encapsulated rod. The bacillus is acquired by droplets, and often it occurs naturally in the respiratory tracts of humans. *Klebsiella pneumonia* may be a primary disease or a secondary disease. As a primary pneumonia, it is characterized by sudden onset and a gelatinous reddish-brown sputum.

The organism grow over the lung surface and rapidly destroy the tissue, often causing death. In its secondary form, *Klebsiella* pneumonia is a *nosocomial*, or hospital-acquired, disease. Bacteria are spread to hospitalized patients by such routes as clothing, intravenous solutions, foods, and the hands of health-care personnel. Usually, the patient is suffering from an already established disease such as emphysema or diabetes, hence the "secondary" connotation. The symptoms, while similar to those in the primary disease, are compounded by the patient's other illness.

### *Serratia* Pneumonia

For many decades, microbiologists considered *Serratia marcescens* a non-pathogenic bacillus and often used it as a test organism in their experiments. Today they view this Gram-negative rod as a cause of nosocomial disease. A patient may be predisposed to infection by such conditions as chronic illness, impaired immunity due to immuno-suppressive therapy, radiation, or surgical treatments such as urinary catheterization, lumbar puncture, or blood transfusion.

*Serratia pneumonia* is accompanied by patches of bronchopneumonia and, in some cases, substantial tissue destruction in the lungs. The major clinical problem in treating the disease is resistance to antibiotic therapy, apparently due to R factors. The literature of the 1990s has also cited *Serratia marcescens* as a cause of eye infections, bone disease, arthritis, and meningitis. In addition, it is an agent of urinary tract diseases.

### Legionnaires' Disease (Legionellosis)

From July 21 to July 24, 1976, the Bellevue-Stratford hotel in Philadelphia was the site of the fifty-eight annual convention of Pennsylvania's chapter of the American Legion. Toward the end of the convention, 140 conventioneers and 72 other people in or near the hotel became ill with fever, coughing, and pneumonia. Eventually, 34 individuals died to the disease or its complications.

Initially some scientists suspected that the outbreak was the first wave of a swine flue epidemic predicted for that year, but as the weeks wore on, it became apparent that no known microorganism was responsible. By early December, the mystery had deepened to the point that a writer from the respected journal *Science* described the "investigation that failed." Finally, in January 1977 investigators from the CDC announced the isolation of a previously unknown bacterium from the lung tissue of one of the patients. The organism appeared responsible not only for the Legionnaires' diseases (as the disease had come to be known) but also for a number of other unresolved pneumonialike diseases. The writer from *Science* acknowledged that "the investigation may be successful after all".

In contemporary microbiology, *Legionnaires' disease* (or *legionellosis*) is known to be caused by the organism isolated in 1977, a Gram-negative rod named *Legionella pneumophila*. The bacillus exists where water collects, and apparently it becomes airborne in wind gusts and breezes. Cooling towers, industrial air-conditioning units, lakes, stagnant pools, and puddles of water have been identified as sources of bacteria. Humans breathe the contaminated droplets into the respiratory tract, and disease develops a few days later.

The symptoms of Legionnaires' disease include fever, a dry cough with little sputum, and some diarrhea and vomiting. In addition, chest X rays show a characteristic pattern of lung involvement. Erythromycin is effective for treatment, and person-to-person transmission is uncommon. In the laboratory, the organisms are difficult to cultivate, but they can be grown on media containing hemoglobin and special amino acid additives.

Once *L. pneumophila* was isolated in early 1977, microbiologists found that the organism was responsible for many epidemics of pneumonia in previous years. One such outbreak, called *Pontiac fever*, took place in Michigan in 1968; another was recorded in Washington, D.C., in 1965. In the years after 1977, reports of Legionnaires' disease occurred throughout the world and, invariably, water was involved. For example, 66 cases in Sweden were linked to water collecting on the rooftop of a shopping center, and 23 cases in Italy were traced to well water used for bathing. In South Dakota in 1990, 26 cases and ten deaths were linked to the water in showers at a particular hospital. Investigators isolated *Legionella pneumophila* from shower heads in all the patients' rooms where illness broke out.

Since *Legionella's* discovery in water, microbiologists have been perplexed as to how such fastidious bacilli could survive in an aquatic environment that is often hostile. An answer was suggested by a 1983 study showing that the bacilli could live and grow within the protective confines of waterborne protozoa. In the South Dakota episode cited above, amebas were found in abundance in the hospital's water supply. The water was treated by heating and by addition of chlorine to help quell the spread of legionellae. Over 1000 cases of Legionnaires' disease were reported to the CDC in 1994, and some health officials estimate that up to 50,000 cases occur annually in the United States. Thus far, six species of *Legionella* have been identified, including *L. pneumophila*.

A host of "legionlike organisms" is presently under investigation as causes of human disease.

**Q Fever**

Q fever is one of several diseases caused by a group of bacteria known as *rickettsiae* (sing., rickettsia). Once regarded as an intermediate type of organism between bacteria and viruses, the rickettsiae are now recognized as "small bacteria" and are classified in the first volume of *Bergey's Manual of Systematic Bacteriology*. They have no flagella, pili, or capsules, and they reproduce by binary fission. With a few exceptions, rickettsiae are cultivated in the laboratory only within living tissue cultures such as fertilized eggs or animals. Arthropods are usually involving in their transmission.

The term *Q fever* was first used by E.H. Derrick in 1937 to describe an illness that broke out among workers at a meat-packing plant in Australia. The "Q" may have been derived either from "query," meaning unknown, or from Queensland, the province in which the disease occurred. The rickettsia that causes Q fever is named *Coxiella burnetii*, after H.R. Cox, who isolated it in Montana in 1935, and Frank Macfarlane Burnet, who studied its properties in the late 1930s.

Q fever is prevalent worldwide among livestock, especially dairy cows, goats, and sheep, and outbreaks may occur wherever these animals are raised, housed, or transported. Transmission among livestock and to humans is accomplished primarily by airborne dust particles as well as by respiratory droplets and by ticks. In addition, humans may acquire the disease by consuming raw milk infected with *C. burnetii* or milk that has been improperly pasteurized. Patients experience severe headache, high fever, a dry cough, and occasionally, lesions on the lung surface. The mortality rate is low, and treatment with tetracycline is effective.

Q fever is an occupational hazard of those who work with domestic animals. Q fever can also be contracted from domestic cats. In 1987 in Nova Scotia, for example, twelve members of a poker group came in contact with the homeowner's infected cat and all contracted the disease. One died of disease-related pneumonia. The organism is among the most heat-resistant of all bacteria, and its elimination is a prime object of milk pasteurization.

**Psittacosis and Chlamydial Pneumonia**

Psittacosis and chlamydial pneumonia are airborne diseases caused by different species of chlamydia. Chlamydiae are a subgroup of rickettsiae and are among the smallest bacteria (0.25 $\mu$m) related to disease in humans. They are cultivated only in living human cells and have a complex life cycle that includes a number of different forms. Three species of chlamydiae are recognized: *Chlamydia trachomatis*, the cause of trachoma and two sexually transmitted diseases; *Chlamydia psittaci*, the cause of psittacosis; and *Chlamydia pneumoniae*, the cause of chlamydial pneumonia.

*Psittacosis* affects parrots, parakeets, canaries, and other members of the psittacine family of birds (*psittakos* is the Greek word for parrot). The disease also occurs in pigeons, chickens, turkeys, and seagulls, and some microbiologists prefer to call it *ornithosis* to reflect the more widespread occurrence. Humans acquire *Chlamydia psittaci* by inhaling airborne dust or dried droppings of infected birds. Sometimes the disease is transmitted by a bite from a bird or via the respiratory droplets from another human. A notable series of cases occurred in 1992 after individuals came in contact with infected parakeets and cockatiels.

The symptoms of psittacosis resemble those of influenza, Q fever, or primary atypical pneumonia. Fever is accompanied by headaches, dry cough, and scattered patches of lung infection. Without antibiotics, the mortality rate outbreaks may reach 20 percent, but tetracycline therapy reduces this number of less than 1 percent. The incidence of psittacosis in the United States is currently very low (about 100 to 250 cases per year) partly because federal law requires a 30-day quarantine for imported psittacine birds. In addition, birds are given chlortetracycline hydrochloride (CTC)-treated water and CTC-impregnated feed. Psittacosis and other diseases of the lower respiratory tract are summarized.

*Chlamydial pneumonia* is caused by *Chlamydia pneumoniae*, an organism formerly called TWAR because two of the original isolates of the organism were specified TW-183 and AR-39. The chlamydia is transmitted by respiratory droplets and causes a mild "walking pneumonia" principally in young adults and college students. The disease is clinically similar to psittacosis and primary atypical pneumonia and is characterized by fever, headache, nonproductive cough, and infection of the lower lobe of the lung. Treatment with tetracycline or erythromycin hastens recovery from the infection. The organism was first observed in Seattle, Washington, in 1983 and is believed to infect many thousands annually in the United States.

5

# FOODBORNE AND WATER-BORNE DISEASES

Poultry products and eggs are well known for their connections to *Salmonella* diseases, but the signs that summer did not point to the poultry and eggs; they pointed to the fruit salad.

There were 75 people at the party in New Jersey that June night of 1991. By the next week, 17 were ill with nausea, vomiting, diarrhea, abdominal cramps, and fever. Soon investigators were visiting each one, asking questions and requesting stool samples. All seventeen had eaten the fruit salad—the watermelon, cantaloupe, honeydew melon, strawberries, and grapes. The investigators scratched their heads because fruit salad and *Salmonella* just don't go together. Or do they? And if they do, which was the responsible fruit? They waited for additional cases to surface.

They did not have to wait long. In July, 20 cases of salmonellosis were reported in Minnesota, and several more occurred in Illinois, Pennsylvania, North Dakota, Missouri, and Michigan. Even Canada contributed 72 cases. All told, there were 400-plus cases of salmonellosis in 23 states and Canada that June-July period. And the culprit seemed to be the cantaloupe.

At least the laboratory researchers suspected it was the cantaloupe because even though they could not isolate *Salmonella* from the fruit, they recovered *Salmonella* from enough patients who had eaten cantaloupe and they were able to pinpoint *Salmonella poona* as the epidemic's cause. In addition, they traced all the cantaloupes to a farm in south Texas where the soil was heavily contaminated by

*Salmonella* and where *S. poona* had probably infected the rinds. But people do not eat cantaloupe rinds; they eat the fruit. So how did the *Salmonella* get from the rind to the fruit? Any guesses?

Salmonellosis will be one of the major foodborne and waterborne diseases we consider in this chapter. Though many diseases such as salmonellosis are rarely life-threatening, others such as botulism, typhoid fever, and cholera can bring an individual to the brink of death if aggressive therapy is not instituted immediately. Public health officials must be constantly vigilant to prevent the transmission of these diseases in the food we eat and the water we drink. Sometimes, however, even their strongest precautions are not good enough, and simply slicing through a piece of fruit can bring bacteria from the rind to the edible portion inside, as probably happened hundreds of times during a certain epidemic in the summer of 1991.

We shall categorize the foodborne and waterborne diseases as intoxications or infections. *Intoxications* are diseases in which bacterial toxins, or poisons, are ingested in food and water. Examples are botulism, staphylococcal food poisoning, and clostridial food poisoning. By contrast, *infections* refer to diseases where live bacteria are ingested in food and water and subsequently grow in the body. These infections include typhoid fever, salmonellosis, shigellosis, cholera, diarrheas, and several other bacterial diseases.

## Foodborne and Waterborne Intoxications

### Botulism

Of all the foodborne intoxications in humans, none is more dangerous than *botulism*. The causative agent, *Clostridium botulinum*, produces a toxin so powerful that one pint of the pure material could eliminate the entire population of the world; one ounce would kill all the people in the United States.

*Clostridium botulinum* is a Gram-positive anaerobic bacillus that forms spores. The spores exist in the intestines of many humans as well as numerous fish, birds, and barnyard animals. They reach the soil in manure, organic fertilizers, and sewage, and often, they cling to harvested products. When spores enter the anaerobic environment of cans or jars, they germinate to bacilli, and the bacilli produce the toxin. The toxin, a protein of high molecular weight (900,000 daltons), is called an *exotoxin* because the bacilli release it to the environment as it is manufactured. Contaminated cans of food usually lack evidence of the toxin. The food tastes and smells normal, and the can is rarely swollen and produces no swish of gas when opened. The bacteria themselves are of little consequence, but the toxin is lethal once absorbed into the bloodstream.

The symptoms of botulism develop within hours. Patients suffer blurred vision, slurred speech, difficulty in swallowing and chewing, and laboured breathing. The limbs lose their tone and become flabby, a condition called *flaccid paralysis*. Often the patients are thought to be suffering from a stroke or Guillain-Barre syndrome or

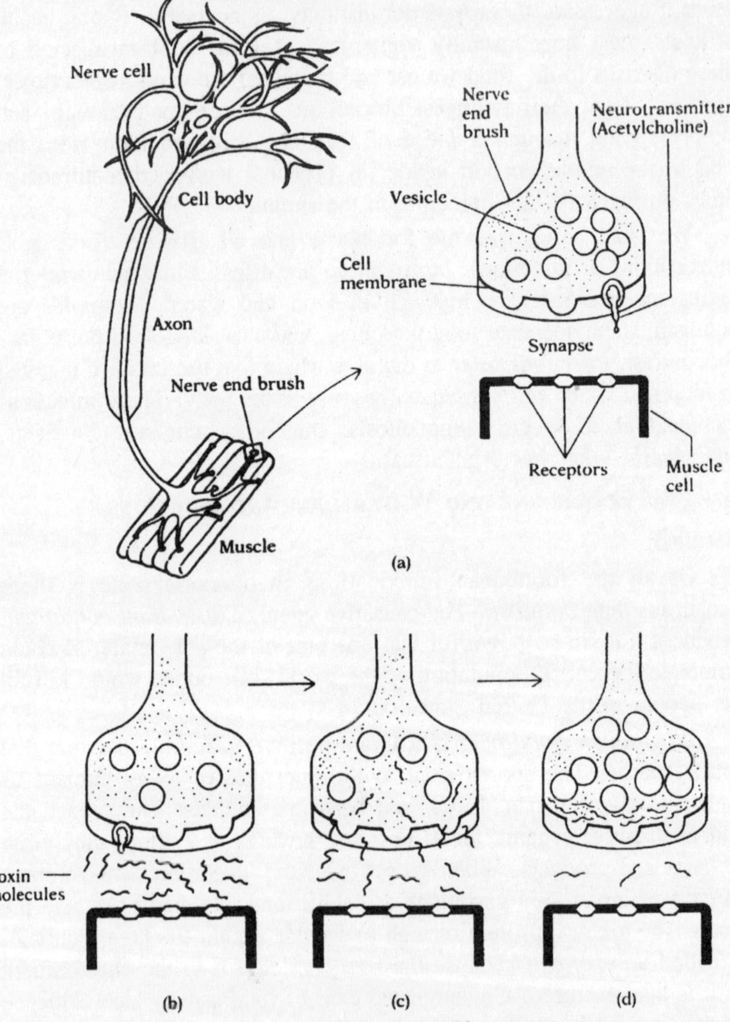


*Fig. 5.1. The mechanism of action of botulism toxin.*

myasthenia gravis. These symptoms result from a complex process in which the toxin penetrates the end brushes of nerve cells and inhibits the release of the neurotransmitter acetylcholine into the junction of nerves and muscles. Without acetylcholine, nerve impulses cannot pass into the muscles, and the muscles do not contract. Failure of the diaphragm and rib muscles to function leads to respiratory paralysis and death within a day or two.

Because botulism is a type of foodborne intoxication (or poisoning), antibiotics are of no value in therapy. Instead, large doses of specific antibodies called *antitoxins* must be administered to neutralize the toxins. Life support systems such as respirators are also utilized. Frequently the nature of the problem is not recognized because the patient has done nothing extraordinary in the previous hours. Identification of the toxin by injecting leftover food samples or patient's serum into laboratory animals often comes too late. Although the annual number of cases in the United States is low, the percentage of deaths is considerable.

Botulism can be avoided by heating foods before consumption, because the toxin is destroyed on exposure to temperatures of 90°C for ten minutes. However, experience shows that most cases are related to food eaten cold. The largest episode recorded in the United Stales, for example, occurred in Michigan in 1977 when 58 people became ill after eating home-canned peppers at a restaurant. Another notable outbreak, involving canned salmon, took place in 1982, and chopped garlic, in oil was apparently the source in a 1985 Canadian incident. Other foods linked to botulism include mushrooms, olives, salami, and sausage. (Indeed, the word "botulism" is derived from the Latin, *botulus*, for sausage.)

Scientists have identified seven types of *Clostridium botulinum* depending on variant of toxin produced. The types are lettered A through G. Types A, B, and E cause most human disease; types C and D cause most animal disease, and type F, most fish disease. Knowing which type of *C. botulinum* caused the disease is important because antitoxin therapy must be type-specific. In animals, botulism is manifested as *fodder disease*, acquired when cattle ingest toxin from silage and feed; in fowl, it is called *limberneck*. Massive fish kills also occur periodically from the consumption of botulism toxin in water.

Studies in recent decades indicate that *Clostridium botulinum* can grow in the anaerobic tissue of a wound and cause *wound botulism*.

Moreover, researchers have implicated the organisms in cases of *infant botulism* and have suggested that the disease may lead to sudden infant death syndrome (SIDS), a condition affecting 8000 infants annually and sometimes called "crib death." In this instance, clostridial spores may germinate in the infant's intestine before other organisms arrive and establish competition. Honey has been implicated as a possible source of the spores. Another species of *Clostridium*, *C. difficile*, may also be the cause.

Before we leave botulism we should note that in recent years, the botulism toxin has been put to use in the service of humanity. Scientists have found that, in minute doses, the toxin can relieve a number of movement disorders (the so-called *dystonias*) caused by involuntary sustained muscle contractions. For example, "botox," as the toxin is known, can be used to treat strabismus, or misalignment of the eyes commonly known as cross-eye. Also, the toxin may be valuable to relieve stuttering, uncontrolled blinking, musician's cramp (the bane of the violinist), and spastic closure of the anal and urinary sphincter. In addition, the toxin could one day become an important treatment for cerebral palsy and multiple sclerosis.

**Staphylococcal Food Poisoning**

Years ago it was common for people to complain of ptomaine poisoning shortly after eating contaminated food. Indeed, the recovery from the intestine of *ptomaines* appeared to justify their reputation. Modern microbiologists, however, have exonerated the ptomaines and placed the blame for most food poisonings on the Gram-positive bacterium *Staphylococcus aureus*. Today, *staphylococcal food poisoning* ranks as the second most reported of all types of foodborne disease, with *Salmonella*-related illnesses ranking first. Because most staphylococcal outbreaks probably go unreported, it may well be that staphylococcal food poisoning is the most common type.

Like botulism, staphylococcal food poisoning is caused by an exotoxin excreted in foods during bacterial growth. Since the symptoms are restricted to the intestine, the toxin is called an *enterotoxin* ("entero" refers to intestine). Patients experience abdominal cramps, nausea, vomiting, prostration, and diarrhea as the toxin encourages the release of water (the word diarrhea is derived from the Greek stems *dia* meaning through and *rhein* meaning to flow; hence water "flows through" the intestine). Sufferers often report feelings of fever, but few actually have fever. The symptoms last for several hours, and recovery is usually rapid and complete.

The *incubation period* for staphylococcal food poisoning is a brief one to six hours. Often the individual can think back and pinpoint the source. Examples are spoiled meats and fish, dairy products, cream-filled pastries, and salads such as potato salad and coleslaw. Foods containing *S. aureus* lack an unusual taste, odor, or appearance, and the only clues to possible contamination are factors such as moisture content, low acidity, and improper heating previous to arrival at the table. The staphylococcal enterotoxin is among the most heat-resistant of all exotoxins.

The main reservoir of *S. aureus* in humans is the nose. Thus an errant sneeze may be the source of staphylococci in foods. Studies indicate, however, that the most common *mode of transmission* is from boils or abscesses on the skin that shed staphylococci. Persons who sell, prepare, or serve food should be particularly alert to these situations. The staphylococci grow over the broad temperature range of 8° C to 45° C and since refrigerator temperatures are generally set at about 5° C, refrigeration is not an absolute safeguard. Ham is particularly susceptible because staphylococci tolerate salt.

*Staphylococcus aureus* normally does not grow in the human intestine because of competition by other organisms. Therefore public health investigators searching for a cause of the food poisoning are usually unable to locate the organisms in stool samples. Moreover, the contaminated food has often been consumed completely. Thus case reports are often based on symptoms, patterns of outbreak, and type of food eaten. When investigators locate staphylococci, they can identify the organisms by growth on mannitol salt agar, Gram staining, and testing with bacteriophages to learn the strain involved. Methods such as these were used in 1989 to relate a multicity outbreak of staphylococcal food poisoning to canned mushrooms imported from China.

**Clostridial Food Poisoning**

Since its recognition in the late 1960s, *clostridial food poisoning* has risen to prominence as the second most common type of reported food poisoning, after staphylococcal food poisoning. The causative organism, *Clostridium perfringens*, is widely known as an agent of gas gangrene. This Gram-positive anaerobic sporeformer commonly contaminates protein-rich foods such as meat, poultry, and beans. If the spores survive the cooking process, they germinate to vegetative cells that produce an enterotoxin within oxygen-free portions of the food. Consumption of the toxin leads to illness. Large numbers of vegetative cells may also be consumed into anaerobic pockets in the large intestine where the enterotoxin is produced.

The incubation period for clostridial food poisoning is about 8 to 14 hours, a factor that distinguishes it from staphylococcal food poisoning. Moderate to severe cramping, abdominal pain, and watery diarrhea are common symptoms, but nausea and vomiting are rare. The enterotoxin encourages the outward movement of water from epithelial cells lining the intestine. Recovery is rapid, often within 24 hours, and therapy is generally unnecessary.

Research studies have suggested that *C. perfringens* spores are present in virtually every carcass of beef, pork, and lamb, and thus the potential for food poisoning is enormous. Outbreaks usually result from failure to refrigerate cooked meats promptly or adequately.

## Foodborne and Waterborne Infections

### Typhoid Fever

*Typhoid fever* is among the classical diseases that have ravaged humans for generations. The disease captured the attention of microbiologists a century ago and was studied by Karl Eberth and Georg Gaffky, both associates of Robert Koch. Eberth observed the causative organism, *Salmonella typhi*, in spleen tissue in 1880, and Gaffky isolated it in pure culture four years later.

*Salmonella typhi*, a Gram-negative rod, displays high *resistance* to environmental conditions outside the body. This factor enhances its ability to remain alive for long periods of time in water, sewage, and certain foods. *S. typhi* causes disease only in humans and is transmitted by the Five Fs: flies, food, fingers, feces, and fomites.

*Salmonella typhi* is acid-resistant, and with the buffering effect of food and beverages, it generally survives passage through the stomach. In the small intestine it invades the tissues, causing deep ulcers and bloody stools, but little diarrhea. Blood invasion follows, and after a few days, the patient experiences mounting fever, lethargy, and delirium ("typhoid" is derived from the Greek *typhos*, for smoke or cloud, a reference to the delirium). The abdomen becomes covered with *rose spots*, an indication that blood is hemorrhaging in the skin. Bowel perforation and gall bladder infection are other possible complications.

Diagnostic tests for typhoid fever include the isolation of *S. typhi* from tissue specimens such as blood and stools. Additional data may be obtained from the *Widal test*, named for Georges Fernand I. Widal who devised it in 1896. In this procedure, the patient is tested (or the

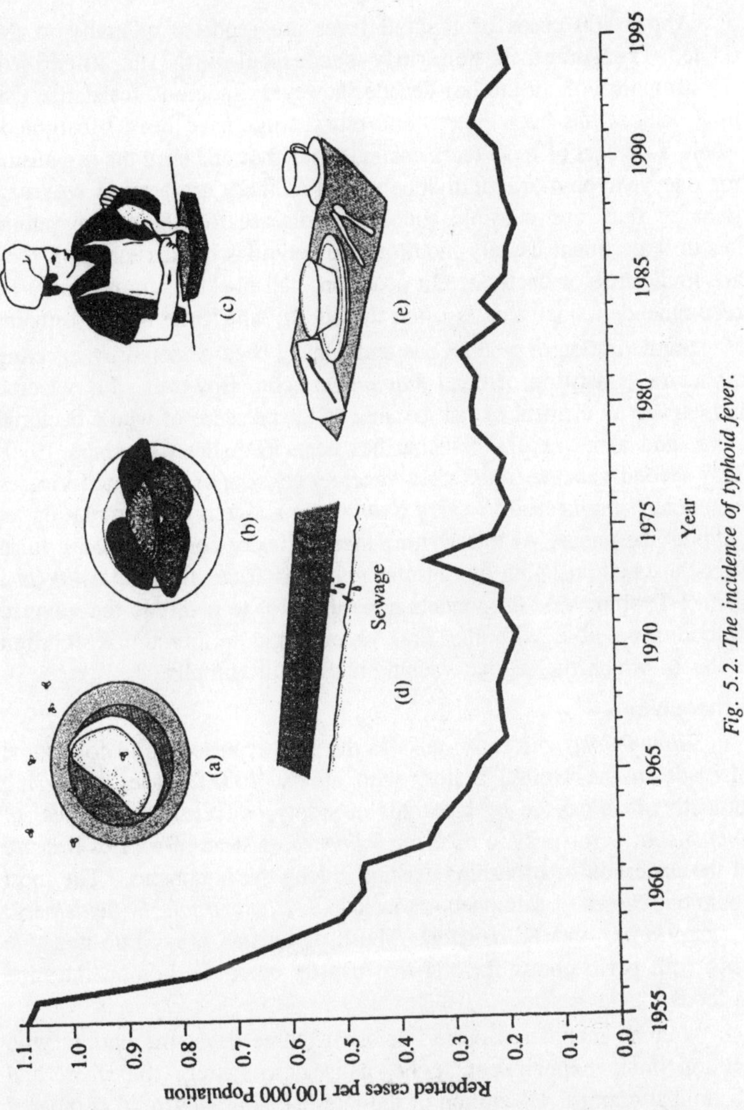


*Fig. 5.2. The incidence of typhoid fever.*

presence of *Salmonella* antibodies by mixing a sample of serum with chemicals derived from *Salmonella* flagella or cells to determine whether a clumping reaction will take place. A third procedure is to ask the patient to swallow a gelatin capsule tied to a string. After several hours, the capsule is retrieved and tested for *S. typhi*. Typing with bacteriophages can also detect the organism.

About 400 cases of typhoid lever are reported annually to the CDC. Treatment is generally successful with the antibiotic chloramphenicol. In the last decade, however, antibiotic resistance due to R factors has been noted, and other drugs have been substituted. About 5 percent of recoverers continue to harbor and shed the organisms for one year or more. Individuals such as these are termed *carriers*. Because they are possible sources of disease to others, the public health department usually monitors the activities of carriers until they are found tree of bacteria. On occasion, gall bladder removal may be recommended since this is often the site in which the bacilli remain.

Immunization to typhoid has traditionally been rendered by injecting a vaccine consisting of dead *Salmonella* cells. However, this vaccine has carried an element of risk because of the presence of whole bacterial cells, and a newer oral vaccine has been FDA-licensed since 1990. This second vaccine (the Ty21a vaccine) consists of weakened viruses genetically engineered lo carry *Salmonella* antigens for stimulating an antibody response. At this writing there is under consideration a third vaccine, one consisting of capsular polysaccharides from *S. typhi* (and named Typhim Vi). Proponents hope to use it to preclude the adverse reactions possible with the first vaccine and to immunize children under 6, which the second vaccine does not accomplish.

**Salmonellosis**

*Salmonellosis* currently ranks as the most reported of all foodborne diseases in the United States, with almost 40,000 cases occurring annually. It is caused by hundreds of serotypes (serological types) of *Salmonella*. Serotypes are used for *Salmonella* instead of species because of the uncertain relationships existing among the organisms. The most common causes of salmonellosis include *S. typhimurium*, *S. heidelberg*, *S. enteritidis*, and *S. newport*. These organisms are Gram-negative rods with peritrichous flagella that usually enter the human intestine in foods.

After an incubation period of one to three days, the patient with salmonellosis experiences fever, nausea, vomiting, diarrhea, and abdominal cramps. Ulceration of the intestine is usually mild compared to that in typhoid lever, and blood invasion is uncommon. The symptoms may last a week or more and generally, the disease is allowed to run its course rather than treat the patient with antibiotics that may damage the tissues. Dehydration may occur in some patients and fluid replacement may be necessary. Diagnosis usually consists of isolating the *Salmonella* serotype from stool specimens or rectal swabs using differential media. Often the infection is called *gastroenteritis* or simply, *enteritis*.

With increased awareness and modern methods of detection, salmonellosis has been linked to a broad variety of foods. Pasteurized milk was linked to 5770 cases in the Midwest in 1985, and frozen pasta products were the source of dozens of cases in the Northeast in 1986. *Poultry products* are particularly notorious because *Salmonella* serotypes commonly infect chickens and turkeys. The organisms may be consumed directly from the poultry or in pot pies, processed chicken roll or turkey roll, or chicken salads. In 1991, the CDC reported an outbreak of salmonellosis in a Connecticut hospital. The outbreak was traced to improperly cooked turkey and was of particular concern because many of the hospital employees were infected and were spreading the disease to already-ill patients. With over 4 billion chickens and turkeys consumed annually in the United States, the possibilities for *Salmonella* contamination are plentiful.

*Eggs* are another widespread source of salmonellosis when used in foods such as custard pies, cream cakes, egg nog, ice cream, and mayonnaise. In 1992 the CDC reported on 38 cases associated with Caesar salad dressing made with raw eggs. In the past, salmonellosis was associated with cracked or contaminated egg shells, bill researchers are *now* coming to believe that *Salmonella* serotypes infect the ovary of the hen and pass into the egg before the shell forms. If this is so, then even the best grade eggs may be contaminated. Consumers should store eggs in the main compartment of the refrigerator, refrigerate leftover egg dishes quickly in small containers (to permit faster cooling), and avoid "runny" or undercooked eggs. The current catchphrase at the CDC is "scramble or gamble." Indeed, in 1992., New Jersey outlawed "eggs over easy." (It later rescinded the ban.)

Live *animals* are also known to transmit salmonellosis, and many states now prohibit the sale of Easter chicks and ducklings for this reason. Moreover, some years ago, the Food and Drug Administration prohibited the distribution of pet turtles less than four inches long because a significant number of salmonellosis outbreaks were attributed to handling these animals. The sale of iguanas is also restricted, and a savannah monitor lizard was apparently the cause of an infant's disease in 1992.

**Shigellosis (Bacterial Dysentery)**

Members of the genus *Shigella* were first described by the Japanese investigator Kiyoshi Shiga in 1898, and two years later by the European microbiologist, Simon Flexner. *Shigella* bacilli are small Gram-negative rods found mainly in humans and other primates. They cause digestive disturbances ranging from mild diarrhea to a severe *and* sometimes fatal dysentery. *Dysentery* is a syndrome manifested by waves of intense abdominal cramps and frequent passage of small-volume, bloody mucoid stools. For many years, *Shigella* diseases were known as *bacterial dysentery*. However, watery diarrhea without blood or mucus is a more common symptom than dysentery, and the name shigellosis is currently preferred.

*Shigellosis* may be caused by any of four serotypes of *Shigella*, including *Sh. sonnei*, *Sh. dysenteriae*, *Sh. flexneri*, and *Sh. boydii*. Of these, *Sh. sonnei* is the most common. Humans can ingest, the organisms in contaminated water as well as in many foods, especially eggs, vegetables, shellfish, and dairy products. *Shigella* usually penetrates the epithelial cells lining the intestine and, after two to three days, it produces sufficient enterotoxins to encourage the cells to release water. Infection of the small intestine produces watery diarrhea, but infection of the large intestine results in the expulsion of bloody mucoid stools characteristic of dysentery. Indeed, the dysentery can be quite indisposing. It is said that at the Battle of Crecy in 1346, the English army was racked with dysentery—when the French attacked, they literally caught the English with their pants down.

Most cases of shigellosis subside within one week without, complications. However, patients who lose excessive fluids must be given salt tablets or oral solutions or intravenous injections of salt solutions for *rehydration*. This treatment is most often applied to infants and debilitated adults in whom circulatory collapse may lead to death. Antibiotics are sometimes effective, but many strains of *Shigella* are resistant due to *R* factors, and the shigellosis may be fatal. Recoverers generally become carriers for a month or more and continue to shed the bacilli in their feces. Identification of the causative agent is confirmed by the isolation of *Shigella* from stool specimens. Immunizing agents are not available, but 586 passengers aboard the cruise ship *Viking Serenade* wish they were. They were the unfortunate victims of a shigellosis outbreak in 1994 traced to bacteria in the water supply. Trapped hopelessly aboard ship, they could only wait for landfall and dream of a more pleasant serenade the next time.

**Cholera**

No diarrhea can compare with the massive diarrhea associated with *cholera*. In the most severe cases, a victim may lose up to one liter of fluid per hour for several hours. The fluid is colourless and watery with characteristic *rice-water stools* reflecting the conversion of the intestinal contents to a thin material like barley soup. The patient's eyes become gray and sink into their orbits. The skin is wrinkled, dry, and cold, and muscular cramps occur in the arms and legs. Despite continuous thirst, sufferers cannot hold fluids. The blood thickens, urine production ceases, and the sluggish blood flow to the brain leads to shock and coma. In untreated cases, the mortality rate may reach 70 percent.

Cholera is caused by *Vibrio cholerae*, a curved Gram-negative rod first isolated by Robert Koch in 1883. The bacilli enter the intestinal tract in contaminated water or food. Vegetables from fields fertilized with nightsoil are a possible source and shellfish such as raw oysters can be involved, as a 1988 case in Colorado illustrated.

*Vibrio cholerae* is extremely susceptible to stomach arid. However, if sufficient numbers are ingested, enough remain to colonize the intestine. As the bacilli move along the intestinal epithelium, they secrete an enterotoxin that stimulates the unrelenting loss of fluid characteristic of the disease. Antibiotics such as tetracycline may be used to kill the bacteria, but the key treatment is replacement of lost fluids and restoration of the water balance in the body. Often, this entails *intravenous injections* of salt solutions and the consumption of glucose and salts. More commonly, all but severely dehydrated adults and children can be treated with oral rehydration solution (ORS), an oral solution of electrolytes that will restore the normal balances in the body.

Cholera has been observed for centuries in human populations, and seven pandemics have been documented. The current pandemic began in 1961 in Indonesia and now involves about 35 countries. A major outbreak occurred in Peru and Ecuador early in 1991 and from there spread throughout the region, accounting for 731,000 cases by the end of 1992 and 6300 deaths in 21 countries of the Western Hemisphere. In 1992, the United States reported 102 cases of cholera, more than in any year since surveillance began in 1961. Seventy-five additional cases occurred in airline passengers arriving in Los Angeles from South America. The passengers had consumed contaminated seafood aboard the plane. The agent, *V. cholerae* biotype El Tor serotype Inaba, has been identified in estuaries along the U.S. Gulf Coast and in ballast waters of ships from South America. By 1994, the epidemic was spreading throughout Africa at what the World Health Organization called a "catastrophic rate."

*Fig. 5.3. An illustration depicting the spread of the cholera epidemic.*

At present, travelers to cholera regions of the world are immunized with preparations of dead *V. cholerae* and receive protection for approximately six months. This may change because public health officials anticipate a genetically engineered cholera vaccine that will give longer-lasting protection without the risk of using dead pathogenic cells. In 1992, researchers at the University of Maryland identified the genes for enterotoxin production and successfully removed the genes from experimental *V. cholerae* cells. So treated, the vibrios became nonpathogenic but still alive (i.e., attenuated), and they could be used in a vaccine to stimulate an antibody response. Field trials for the new vaccine are ongoing at this writing. In the meantime, the most important preventive measures are sanitation, personal hygiene, and care in food preparation. Outbreaks in the United States are rare, but the disease is estimated to cause an annual four to five million deaths throughout the world.

### *E. coli* Diarrheas

One of the major causes of *infantile diarrhea* is the Gram-negative rod *Escherichia coli*. This organism may induce diarrhea by either of two mechanisms: certain "enterotoxic" strains produce an enterotoxin similar to that in cholera, while other "enteroinvasive" strains penetrate the intestinal epithelium as in shigellosis. The toxin causes fluid loss in the small intestine, while the penetration occurs primarily in the large intestine. Both mechanisms lead to dehydration and salt imbalance substantial enough to be life-threatening in infants. Vigorous antibiotic therapy and fluid replacement are usually effective. Often the infection is a nosocomial infection.

*Traveler's diarrhea* is a term usually applied to a disease in which the victim experiences diarrhea within two weeks of traveling to a tropical location; the diarrhea lasts one to ten days. Sometimes this disease is called *Montezurna's revenge*, a reference to the leader of the Aztec nation decimated by smallpox and other diseases brought from Europe in the 1500s. A number of organisms including several bacteria, viruses, and protozoa may cause traveler's diarrhea, but several recent studies point to *E. coli* as the principal agent. The bacilli adhere to the intestinal lining with pili and produce enterotoxins, which induce water loss. The volume lost is usually low, but occasionally it may be considerable, and dysentery may occur. The possibility of traveler's diarrhea may be reduced by careful hygiene and attention to the food and water consumed during visits to other countries.

In recent years, it has become apparent that one can contract a rather serious *E. coli* diarrhea without traveling. In 1993, for example, *E. coli* 0157:H7, an enterotoxic strain, was the cause of over 500 cases of serious illness in patrons of the Jack-In-The-Box fast-food chain. Also that year, mayonnaise contaminated with *E. coli* was implicated in numerous cases of bloody diarrhea contracted at Sizzler restaurants in Oregon; and some years ago the McDonald's chain had to face an outbreak of *E. coli* infection traced to its hamburgers. In 1991, *E. coli* 0157:H7 was seemingly responsible for 70 cases of illness among attendees at an agricultural threshing show. Statistical analyses showed that contaminated roast beef served at the dinner was the most likely source of bacteria. Slices of the meat were served from warm roasting pans probably of sufficient temperature to encourage bacterial growth.

Epidemiologists at the CDC now estimate that approximately 20,000 cases of *E. coli* 0157:H7 infection occur annually in the United States, with about 250 deaths. Forty-six clusters of infections were recognized in 1994. As of 1995, 32 states required that *E. coli* 0157:H7 isolates be reported to the state health department, and physicians were alerted to watch for cases of bloody diarrhea. A toxin similar to that produced by *Shigella* species has been identified in *E. coli* isolates (it is called the "shigalike toxin"). Fresh-picked carrots in a garden salad were implicated in two incidents in 1993, and dry-cured salami was implicated in two 1994 outbreaks.

**Peptic Ulcer Disease**

One of the more remarkable discoveries of the modern era is that many cases of peptic (stomach) ulcers are caused by a bacterium. Traditionally, ulcers have resulted from "excess acid" due to factors such as nervous stress, smoking, alcohol consumption, diet, and physiological dysfunction. However, the work of Barry Marshall and his co-workers has made it clear that the bacterium *Helicobacter pylori* is involved. This Gram-negative curved rod is similar to *Campylobacter* species and is apparently transmitted by contaminated water. Doctors have revolutionized the treatment of ulcers by prescribing antibiotics such as tetracycline and have achieved cure rates of up to 90 percent; and relapses are uncommon.

How *H. pylori* manages to survive in the intense acidity of the stomach is interesting. Apparently, the bacterium twists its way through the mucous coaling of the stomach lining and attaches to the stomach wall. There it secretes the enzyme urease. Urease digests urea in the area and produces ammonia as an end product. The ammonia neutralizes acid in the stomach, and the organism begins its destruction of the tissue supplemented by digestive enzymes normally found in the stomach tissue. Soon a sore appears in the stomach lining of perhaps one-quarter to one-half inch diameter (although some ulcers may be up to an inch in diameter). The pain is severe and is not relieved by food or an antacid (as is a duodenal ulcer).

The association of peptic ulcers with a bacterium has brought permanent relief to thousands of sufferers worldwide. But the research is not yet done because the modes of transmission must be studied, the treatments await refining, and the organism's relationship to stomach cancer remains to be elucidated. Still, these are historic times for patients and physicians alike.

**Campylobacteriosis**

Since the early 1970s, *campylobacteriosis* has emerged from being an obscure disease in animals to recognition as a widespread intestinal disease in humans. The causative organism is *Campylobacter jejuni*, a curved (*campylo*-means "curved"). Gram-negative rod that moves by means of a single polar flagellum. Reservoirs for the organism include the intestinal tracts of many animals, including dairy cattle, chickens, and turkeys. Contaminated water is also a source of infection.

The clinical symptoms of campylobacteriosis range from mild diarrhea to severe gastrointestinal distress with fever, abdominal pains, and bloody stools. *Campylobacter* colonizes the small or large intestine, causing inflammation and occasional mild ulceration. However, the signs and symptoms of campylobacteriosis are not unique. Most patients recover in less than a week without therapy, but some have high fevers and bloody stools for prolonged periods. Erythromycin therapy hastens recovery.

In 1982, eleven states reported a total of 3900 isolations of *Campylobacteria jejuni*, and in two states, the number of isolations exceeded that for *Salmonella*. Raw milk was identified as a source of *Campylobacter* in 18 outbreaks between 1981 and 1990, most outbreaks involving school field trips to local dairy farms. A notable incident in New Zealand in 1990 occurred at a camp where contaminated water was apparently the source of the *Campylobacter*. Other species, including *C. coli* and *C. upsaliensis*, have also been identified as human pathogens, and as methods of detection improve, reports of campylobacteriosis will probably continue to rise.

**Brucellosis**

*Brucellosis* is an occupational hazard of farmers, veterinarians, dairy and meat plant workers, and others who work with large animals. Transmission can occur by splashing milk into the eye, by accidental passage of fluids through skin abrasions, by contact with animals, and by consumption of milk and other dairy products. In one example of food transmission, the CDC reported 29 cases in Mexican emigrants living in Houston, Texas. All had eaten goat cheese made from unpasteurized goat's milk and sold by street vendors. Human-to-human transmission is virtually unknown.

Among the organisms responsible for brucellosis are *Brucella abortus* from cattle, *B. suis* from swine, and *B. mellitensis* from goats and sheep. Another species, *B. canis*, is associated with dogs and may be acquired by contact with pets. All are small Gram-negative rods. In animals, brucellosis manifests itself in several organs, especially those of reproduction since it multiplies in the tissues of the uterus. Sterility is a common complication, and pregnant animals with the

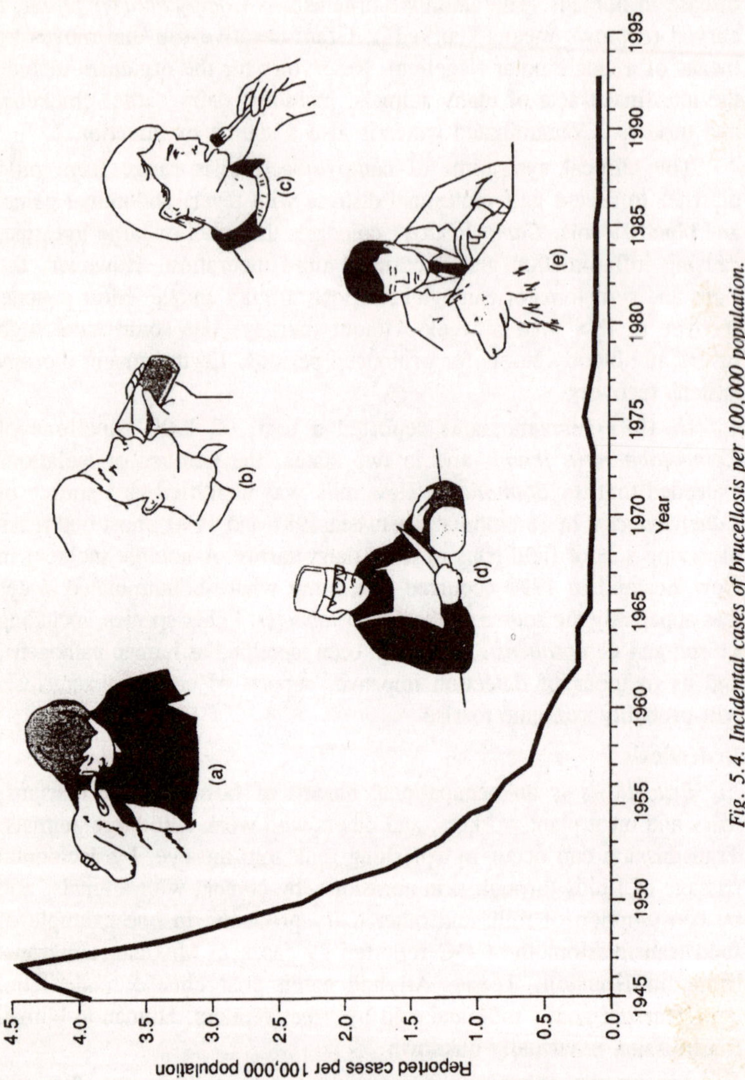


*Fig. 5.4. Incidental cases of brucellosis per 100,000 population.*

disease are known to abort their young. In veterinary literature, the disease is often referred to as *contagious abortion*.

The major focus of brucellosis in humans is the blood-rich organs such as the spleen and lymph glands. Patients experience influenzalike weakness as well as backache, joint pain, and a lever that is high with drenching sweats in the daytime and low with chills in the evening. This fever pattern gives the disease its alternate name of *undulant fever* (*undulat*-is Latin for "vary"). The disease seldom causes death in humans, and tetracycline speeds recovery. Pasteurization of milk and livestock immunization have reduced the incidence of brucellosis in humans from 6300 cases in 1947 to about 100 cases annually.

Much of the early work on brucellosis was performed by David Bruce, after whom the organism is named. Bernard L.F. Bang studied the disease in cows in the 1890s and for a while it was known as *Bang's disease*. Another name, *Malta fever*, was derived from an outbreak among soldiers on the island of Malta in the 1880s. Alice Evans later showed that these various diseases were all related to one another.

## Other Foodborne and Waterborne Diseases

A number of other bacterial organisms transmitted by food and water merit consideration in this chapter. Some have been recognized as pathogens only in recent years; others have been known for decades.

Among the recent, concerns is *Vibrio purahaemolyticus*. This Gram-negative rod is a major cause of foodborne infections in Japan and other areas of the world where seafood is the main staple of the diet. Patients experience acute abdominal pain, vomiting, diarrhea, and watery stools. Some years ago, an outbreak aboard an American cruise ship was linked lo seafood salad, and another incident in Louisiana was traced to unrefrigerated cooked shrimp.

*Bacillus cereus* is a Gram-positive sporeforming bacillus that causes food poisoning in two distinct forms, both due to enterotoxins. The first form ("diarrheal") is accompanied by diarrhea and abdominal pain, while the second form ("emetic") is characterized by substantial vomiting, frequently experienced after consuming cooked rice. Neither form involves lever, and most patients recover within two days without therapy. In the 1980s, investigators traced a notable outbreak to a college cafeteria where a macaroni and cheese dish harbored a million bacilli per gram. Heat-resistant spores had apparently survived the cooking process. Fried rice prepared at a local restaurant was the cause of a 1993 outbreak at a Virginia day-care center.

Another emerging cause of foodborne illness is *Yersinia enterocolitica*. This Gram-negative rod is widely distributed in animals and in river and lake water. The largest episode of disease thus far recorded in the United States occurred in 1982 and involved 172 patients in Arkansas, Tennessee, and Mississippi. Patients experienced fever, diarrhea, and abdominal pain, and most required hospitalization. Investigators believed that pasteurized milk was the source. A 1990 outbreak in Georgia involved raw chitterlings (pork intestines). In this case, *Y.. enterocolitica* was apparently transferred to a group of children by hand-to-hand contact with an individual preparing the food for consumption.

Still another cause of intestinal illness is *Plesiomonas shigelloides*, a Gram-negative, facultatively anaerobic rod. *P. shigelloides* is normally limited to the tropical and subtropical climates of Asia and Africa, and it is commonly found in the gut of tropical fish. It can cause intestinal illness in people who eat raw seafood or who travel in the tropics and consume contaminated water or eat contaminated food. A 1990 case, however, appeared to fit neither description. The case occurred in a year-old child who had apparently been infected by water in the bathroom tub. Her parents previously had poured water from their aquarium into the tub. The aquarium was home to a collection of piranhas, a type of tropical fish the family kept as pets.

Clinical reports in the 1980s have documented that *Aeromonas* species, especially *Aeromonas hydrophila*, cause human gastrointestinal disease. The organisms are Gram-negative rods commonly found in soil and water. They appear to be transmitted by food. Both choleralike and dysenterylike diarrheas, ranging from mild to serve, have been reported in patients. An enterotoxin may be responsible for the symptoms.

6

# SOILBORNE DISEASES

The Black Death (the "plague") was probably the greatest catastrophe ever to strike Europe. It swept back and forth across the continent for almost a decade, each year increasing in ferocity. By 1348, two thirds of the European population was stricken and half of the sick and died. Houses were left empty, towns were abandoned, and a dreadful solitude hung over Europe. The sick died too fast for the living to bury them, and at one point, the Rhone River was consecrated as a graveyard for plague victims. Contemporary historians wrote that posterity would not believe such things could happen, because those who saw them were themselves appalled. To many people, it was the end of the world.

Before the century concluded, the Black Death visited Europe at least five more times in periodic reigns of terror. During one epidemic in Paris, an estimated 800 people died each day; in Siena, the population dropped from 42,000 to 15,000; and in Florence, 75 percent of the citizenry is believed to have perished. Flight was the chief recourse for people who could afford it, but ironically, the escaping travelers spread the disease. Those who remained in the cities were locked in their homes until they succumbed or recovered.

The immediate effect of the plague was a general paralysis in Europe. Trade ceased and wars stopped. Peasants who survived encountered unexpected prosperity because landowners had to pay higher wages to obtain help. Land values declined and class relationships were upset as the system of feudalism gradually crumbled. The authority of the clergy, already in decline, deteriorated further because the Church was helpless in the face of the disaster. With many priests and monks dead, a new order of reformers arose to found Protestantism. Medical practices became increasingly sophisticated, with new strands of sanitation and a 40 day period of detention (a "quarantine") imposed on vessels docking at ports. Even the mechanical clock came into widespread use, reflecting the urgency of life.

The graveyard of plague left fertile earth for the renewal of Europe during the Renaissance. To many historians, the Black Death remains a major turning point in Western civilization.

Nor did it end there. European populations were also devastated by typhus and relapsing fever. Both of these diseases, like plague, and transmitted by arthropods and both can be interrupted by arthropod control. Neither is a major problem in our society, but other arthropodborne diseases such as Lyme disease, tularemia, and Rocky Mountain spotted fever occasionally are reported in the news media. We shall study each of these diseases in this chapter.

To begin, we shall examine a number of soilborne diseases where organisms enter the body through a cut, wound, or abrasion, or by inhalation. Among these are anthrax, a feared disease in agriculture, and tetanus, a concern to anyone who has stepped on a nail or piece of glass. We shall also study leptospirosis and listeriosis, two diseases that are receiving wider recognition as detection methods improve. These soilborne diseases as well as the arthropodborne diseases are primarily problems of the blood. Thus you will note the general patterns of blood entry and a series of symptoms that reflect involvement of the cardiovascular system.

## Soilborne Bacterial Diseases

### Anthrax

*Anthrax* is primarily a disease of large animals such as cattle, sheep, and goats. It is caused by *Bacillus anthracis*, a Gram-positive sporeforming rod. Animals ingest the spores from the soil during grazing and are soon overwhelmed with bacteria. Their organs fill with bloody black fluid (*anthrac* is the Greek word for coal; the disease name is thus a reference to the blackening of the blood). About 80 percent of untreated animals die, and since the bacteria remain in the dead body as spores, it is often necessary to cremate the carcass or bury it deeply in lime to prevent soil contamination. Burning the field may also be required.

Humans acquire anthrax in a number of ways. Workers who tan hides, shear sheep, or process wool may inhale the spores and contract pulmonary anthrax, often called *woolsorter's disease*. Consumption of contaminated meat may lead to gastrointestinal anthrax. Contact with spores may lead to anthrax of the skin. Animal products related to outbreaks of anthrax in the past have included violin bows, shaving bristles, goatskin drums, and leather jackets.

Anthrax spores germinate rapidly on contact with human tissues. The thick capsule of the cells impedes phagocytosis, and the organisms produce a number of toxins. Pulmonary anthrax develops into a severe blood infection with extensive hemorrhaging. Skin and intestinal anthrax are accompanied by boillike lesions covered with black crusts. Blood invasion follows, and violent dysentery with bloody stools accompanies the intestinal form. Penicillin is used for therapy.

In the 1980s, the total number of anthrax cases in the United States was less than a dozen, in large measure due to the testing of imported animal products. A vaccine prepared with an attenuated strain of *B. anthracis* has been successful in reducing outbreaks in domestic herds. However, the spores may remain in many soils, and the disease is considered a threat in biological warfare.

**Tetanus**

*Tetanus* is among the most dangerous of human diseases. *Clostridium tetani*, the bacillus that causes it, occurs everywhere in the environment, especially the soil. Spores enter a wound in very small numbers and revert to vegetative bacilli that produce the second most powerful toxin known to science (after botulism). The toxin provokes sustained and uncontrolled contractions of the muscles (tetanus), and spasms occur throughout the body. Patients often experience violent deaths.

*Clostridium tetani* was first isolated in 1889 by the Japanese bacteriologist Shibasaburo Kitasato. It is a Gram-positive anaerobic sporeforming bacillus found in the intestines of many animals and humans. The bacillus possesses few invasive tendencies and does not cause intestinal disease. However, the spores are excreted in the feces to the soil and later enter the dead oxygen-free tissue of a wound. The wound may result from a fracture, gunshot, dog bite, or puncture by a piece of glass, thorn, or needle. Rusty nails pose a threat because spores cling to the rough edges of the nail and because the nail causes extensive tissue damage as it penetrates. (The rust itself is of no consequence.)

Once inside the tissue, the spores germinate to vegetative cells that produce several toxins. The most important of these toxins appears to be *tetanospasmin*, an exotoxin of high molecular weight. At the synapse, this toxin is believed to inhibit the removal of acetylcholine by interfering with the activity of cholinesterase, the enzyme that normally breaks down acetylcholine by interfering with the activity of cholinesterase, the enzyme that normally breaks down acetylcholine. With acetylcholine left in place at the synapse, volleys of spontaneous impulses arise in the nerves and crash into the muscles, causing the muscles to contract. Another viable theory is that the toxin blocks the release of transmitter substances, which normally inhibit nerve impulse transfer and severe as controlling devices. With the inhibitor substances not present, nerve impulses enter the muscle fibers and cause uncontrolled contractions.

Symptoms of tetanus develop rapidly, often within hours. A patient experiences generalized muscle stiffness, especially in the facial and swallowing muscles. Spasms of the jaw muscles cause the teeth to clench and a condition called *lockjaw*. Severe cases are characterized by a "fixed smile," arching of the back, spasmodic inhalation, and seizures in the diaphragm and rib muscles leading to reduced ventilation and death. Patients are treated with sedatives and muscle relaxants, and are placed in quiet, dark rooms. Physicians prescribe penicillin to destroy the organisms and tetanus antitoxin to neutralize the toxin.

Immunization to tetanus may be rendered by injections of tetanus *toxoid* in the diphtheria-pertussis-tetanus (DPT) vaccine. The toxoid, developed in 1933 by Gaston Roman, is prepared by treating the toxin with formaldehyde to eliminate its toxic quality. Children usually receive the first of several injections at the age of two months. Booster injections of tetanus toxoid in the Td vaccine (a "tetanus shot") are recommended every ten years to keep the level of immunity high.

The United States has experienced a steady decline in the incidence of tetanus, with approximately 150 cases confirmed annually. Most cases occur in the young, who are more likely to have contact with soil, and in the elderly, in whom antibody levels have dropped to vaccination nerve took place. In other parts of the world, tetanus remains a major health problem and is often related to traditional customs. In some countries, for example, unsanitary ear-piercing is common, tattooing is widespread, and the umbilical stump of newborns is dressed with soil. In other cases, a simple splinter of wood may be the source of entry.

**Gas Gangrene**

To understand gas gangrene as an infectious disease, it is important to understand the physiological term "gangrene." Gangrene is a condition that develops when the blood flow ceases to a part of the body, usually as a result of blockage by dead tissue. The body part, generally an extremity, becomes dry and shrunken, and the skin colour changes to purplish or black. The gangrene may spread as enzymes from broken cells destroy other cells, and the tissue may have to be excised ("debrided") or the part amputated. This form of gangrene is called "dry gangrene."

*Gas gangrene*, or "moist gangrene," occurs when soilborne bacteria invade the dead anaerobic tissue. The organisms responsible are *Clostridium perfringens* (formerly called *Clostridium welchii*), as well as several other species of *Clostridium* found in the intestine and obtained from the soil in a wound. These anaerobic sporeforming Gram-positive rods multiply rapidly, while fermenting the muscle carbohydrates and putrefying the muscle proteins. Large amounts of gas resulting from this metabolism tar the tissue apart. The gas also presses against blood vessels, thereby blocking the flow and forcing cells away from their blood supply. In addition, the organisms secrete *lecithinase*, an enzyme that dissolves cell membranes and releases toxic cellular enzymes. Two other bacterial enzymes, *hyaluronidase* and *hemolysin*, facilitate the passage of bacteria among the cells and destroy red blood cells, respectively. Neurotoxins are also released.

The symptoms of gas gangrene include intense pain and swelling at the wound site as well as a foul odour. Initially the site turns dull red, then green and finally blue-black. Anemia is common, and bacterial toxins may damage the heart and nervous system. Treatment consists of antibiotic therapy as well as debridement, amputation, or exposure in a hyperbaric oxygen chamber. The disease spreads rapidly, and death frequently results. Some microbiologists prefer the name *clostridial myonecrosis* ("muscle-cell-death") to gas gangrene, because gas is not always present in the early stages of disease.

**Leptospirosis**

*Leptospirosis* is a typical *zoonosis*, that is, a disease of animals able to spread to humans. The disease affects household pets such as dogs and cats, as well as rats, mice, and barnyard animals. Humans acquire it by contact with these animals or from soil or water contaminated with their urine.

The agent of leptospirosis is *Leptospira interrogans*, a small, delicate spirochete usually possessing a hook at one end that resembles a question mark, hence the name "interrogans". The undulating movements of these organisms result from contractions of submicroscopic fibers called *axial filaments*. In animals, the spirochetes colonize the kidney tubules and are excreted in the urine to the soil. They enter the human body through the skin, especially through abrasions and the soft part of the feet. Following penetration, they infect various organs including the kidney, liver, and meninges. Considerable jaundice may be present as bile seeps from the liver, and the patient may vomit blood from gastric hemorrhages. Despite the numerous tissues involved, the mortality rate from leptospirosis is low, and penicillin is generally used with success.

Leptospirosis occurs throughout the United States, and recognition of its symptoms together with modern methods of diagnosis have led to increase reports of its occurrence. The disease is considered as occupational hazard of veterinarians as well as tunnel diggers, sugarcane cutters, dock workers, miners, and others who work in wet areas where rodents are present. Dogs may be immunized with a leptospirosis vaccine usually combined with rabies, parvovirus, and distemper vaccines.

**Listeriosis**

*Listeriosis* is caused by *Listeria monocytogenes*, a small Gram-positive rod. At one time, this organism was dubbed the "Cinderella" of pathogenic bacteria because its ability to cause disease was not widely recognized. The bacillus is because its ability to cause disease was not widely recognized. The bacillus is commonly found in the soil and in many animals, including barnyard animals, dairy cattle, and household pets. It is transmitted to humans by contact with the soil as well as by consumption of animal foods. In 1992 the CDC reported on outbreaks of listeriosis over a previous three-year period and indicated that soft cheeses (e.g., Brie, Camembert, feta, and blue-veined cheeses) as well as delicatessen cold cuts were associated with a significant number of cases.

Listeriosis occurs in many forms. One form, called *listeric meningitis*, is characterized by headaches, stiff neck, delirium, and coma. Another form is a blood disease accompanied by high numbers of white blood cells called monocytes (hence the organism's name). A third form is characterized by infection of the uterus with vague influenzalike symptoms. If contracted during pregnancy, the disease may result in miscarriage of the fetus or mental damage in the newborn.

Listeriosis does not appear to be transmissible among humans. Prompt and prolonged treatment with antibiotics such as tetracycline is effective, but relapses are common. CDC epidemiologists estimate that 2000 cases of listeriosis occur in the United States annually and that a quarter result in the patient's death.

**Melioidosis**

*Melioidosis* is a disease that, until recently, has been rare in the United States. It is caused by a Gram-negative rod named *Pseudomonas pseudomallei*. This organism is common in soil and water, especially in Southeast Asia. Humans contract melioidosis by absorbing bacteria through soil-contaminated wounds, by inhalation, or by consuming contaminated food or water. The disease also affects horses, rodents, dogs, and cats.

Melioidosis has been called a "medical time bomb" for its propensity to lie dormant in the body for years. Abscesses may eventually occur in the heart, lungs, liver, or spleen, and symptoms may vary widely, depending on which organs are affected. Because melioidosis can mimic diseases like pneumonia, arthritis, or heart attack, it is not often recognized. Tropical experts have known of melioidosis since the early 1900s, but Americans did not appear to be involved until the Vietnam War, when several cases surfaced.

## Arthropodborne Bacterial Diseases

**Bubonic Plague**

Few diseases have had the rich and terrifying history of *bubonic plague*, nor can any match the array of social, economic, and religious changes wrought by this disease.

The first documented pandemic of plague probably began in northeast Africa during the reign of the Roman emperor Justinian in 542 A.D. It lasted for 60 years, killed millions, and contributed to the downfall of Rome. The second pandemic was known as the *Black Death* because of the purplish-black splotches on victims and the terror it evoked in the 1300s. The Black Death decimated the world and, by some accounts, killed an estimated 40 million people in Europe, almost a third of the population of that continent. When it was over, feudalism was finished, Protestantism was on the rise, and the Renaissance was ready to flower. A deadly epidemic occurred in London in 1665, where 70,000 people succumbed to the disease. Daniel Defoe's *Journal of the Plague Year* recounts how people reacted.

In the late 1800s, Asian warfare facilitated the spread of a Burmese focus of plague, and migrations brought infected individuals to China and Hong Kong. During this epidemic in 1894, the causative organism was isolated by Alexandre Yersin and, independently, by Shibasaburo

Kitasato. Plague first appeared in the United States in San Francisco in 1900, carried by rats on ships from the Orient. The disease spread to ground squirrels, prairie dogs, and other wild rodents, and it is endemic in the southwestern states today, where it is commonly called *sylvatic plague*.

Bubonic plague is caused by *Yersinia pestis* (formerly *Pasteurella pestis*), a Gram-negative rod. This bacillus stains heavily at the poles of the cell, giving it a safety-pin appearance and a characteristic called *bipolar staining*. As first demonstrated by Masaki Ogata in 1897, the bacillus is transmitted by the rat flea *Xenopsylla cheopis*. A living organism such as a flea that transmits disease agents is called a *vector*, from the Latin *vehere*, meaning "to carry." Normally the fleas infest only rats, but as the rats die, the fleas jump to humans and feed in the skin, thus transferring the bacilli into the bloodstream.

Bubonic plague is basically a blood disease. The bacteria multiply in the bloodstream and localize in the lymph nodes, especially those of the armpits, neck, and groin. Hemorrhaging in the lymph nodes causes substantial swellings called *buboes* (hence, bubonic plague). Dark, purplish splotches from hemorrhages can also be seen through the skin (the "rosies" in "Ring-a-ring of rosies"). From the buboes, the bacilli spread to the bloodstream, where they cause a form of the disease called *septicemic plague*. Over 50 percent of untreated cases of bubonic or septicemic plague are fatal.

Human-to-human transmission of plague during epidemics has little to do with fleas. Instead, it is dependent upon respiratory droplets because *Y. pestis* localizes in the lungs of many of the first victims. In this form, the disease is called *pneumonic plague* and is highly contagious. Lung symptoms are similar to those in pneumonia, with extensive coughing and sneezing. Hemorrhaging and fluid accumulation are common. Many suffer cardiovascular collapse, and death is common within 48 hours of onset of symptoms. Mortality rates for pneumonic plague approach 100 percent. One 1992 death was linked to a cat and rodents and a notable outbreak appears to have occurred in India in 1994.

Plague may be treated with tetracycline or streptomycin when cases are detected early. Diagnosis consists of the laboratory isolation of *Y. pestis* together with tests for plague antibodies and typing with bacteriophages. The disease occurs sporadically in Native American populations and in travelers through the Southwest. Small game hunters, taxidermists, veterinarians, zoologists, and others who have occasion to handle small rodents should be aware of the possibility of contracting plague. About two dozen cases are reported to the CDC annually. A vaccine consisting of dead *Y. pestis* cells is available for high-risk groups.

**Tularemia**

*Tularemia* is one of several microbial diseases first recorded in the United States (others include St. encephalitis, Rocky Mountain spotted fever, Lyme disease, and Legionnaires' disease). In 1911, a U.S. Public Health Service investigator named George W. McCoy reported the plaguelike disease in ground squirrels from Tulare County, California, and within a year he isolated the responsible bacillus. By 1920, researchers identified tularemia in humans, and Edward Francis assumed a detailed study of the disease. Over the next quarter century, Francis amassed data on over 10,000 cases, including his own. In the 1960s, when a name was coined for the causative organism, *Francisella tularensis* was selected to honour him.

*Francisella tularensis* is a small Gram-negative rod that displays bipolar staining. It occurs in a broad variety of wild animals, especially rodents, and it is particularly prevalent in rabbits. Cats and dogs may acquire the bacillus during romps in the woods, and humans are infected by arthropods from the fur of animals. Ticks are important vectors in this regard, as evidenced by an outbreak of 20 tickborne cases in South Dakota in the 1980s. Other methods of transmission include contact with an infected animal (such as skinning an animal), consumption of rabbit meat, splashing bacilli into the eye, and inhaling bacilli, even from laboratory specimens.

Various forms of tularemia exist, depending on where the bacilli enter the body. An arthropod bite, for example, may lead to a skin ulcer and swollen salivary glands; splashing into the eye may cause an eye lesion; and inhalation may lead to pulmonary symptoms. In all forms, the disease is difficult to recognize because the symptoms are mild and nonspecific. Various patients with tularemia have been mistakenly treated for strep throat, rickettsial disease, lymphoid cancer, and cat scratch fever. Diagnosis depends on tests for tularemia antibodies and isolation of *F. tularensis* from body tissues.

Tularemia usually resolves on treatment with tetracycline or streptomycin, and few people die of the disease. Epidemics are unknown, and evidence suggests that tularemia may not be communicable among humans despite the many modes of entry to the body. Physicians report about 200 cases annually in the United States, most from the Midwest.

**Lyme Disease**

One of the major emerging problems of the contemporary era is Lyme disease. Lyme disease is currently the most common tickborne (indeed, arthropodborne) illness in the United States. It has been reported in 47 states, and in 1995, it accounted for over 10,000 cases of infectious disease in the United States.

*Lyme disease* is named for Old Lyme, Connecticut, the suburban community where a cluster of cases was observed in 1975. That year, researchers led by Allan C. Steere of nearby Yale University traced the disease to ticks of the genus *Ixodes*. However, they were unable to locate a causative agent. Several years would pass until 1982, when a spirochete was observed in diseased tissue by investigators at the State University of New York at Stony Brook. Still another two years passed before the spirochete was isolated and cultivated. Researchers named it *Borrelia burgdorferi* for Willy Burgdorfer, the microbiologist who had studied the spirochete in the gut of an infected tick.

Lyme disease has a variable incubation period that can be as long as six to eight weeks. One of the first signs to appear is a slowly expanding red rash at the site of the tick bite. The rash is called *erythema* (red) *chronicum* (persistent) *migrans* (expanding), or ECM. Beginning as a small flat or raised lesion, the rash increases in diameter in a circular pattern over a period of weeks, sometimes reaching a diameter of 10 to 15 inches. It has an intense red border and a red center, and it resembles a bull's eye (the "bull's eye rash"). It can vary in shape and is usually hot to the touch, but it need not be present in all cases of disease. Indeed, about one-third of patients do not develop ECM. The initial tick bite can be distinguished from a mosquito bite because the latter itches, while a tick bite does not. Fever, aches and pains, and flulike symptoms usually accompany the rash.

In the rash (or "early localized") stage of Lyme disease, effective treatment can be rendered with penicillin and tetracycline. Left untreated, some cases resolve spontaneously, but others enter a second stage ("early disseminated") where the disease spreads. Weeks to months later, the patient may experience pain, swelling, and *arthritis* in the large joints, especially the knee, shoulder, ankle, and elbow joints. Multiple rash sites may also be present at this stage reflecting spread of the spirochete. Vigorous treatment at this stage may effectively eliminate the spirochete, but the joint damage may be irreversible.

In the third stage ("late chronic"), the arthritis may be complicated by damage to the cardiovascular and nervous systems. Patients often display irregular heartbeats, migraine headaches, hearing and vision abnormalities, and loss of muscle tone especially in the facial muscles (Bell's palsy). Although Lyme disease is not known to have a high mortality rate, the overall damage to the body can be substantial.

The tick that transmits most cases of Lyme disease in the Northeast and Midwest is *Ixodes scapularis* (formerly *I. dammini*); in the West the major vector is *I. pacificus*. In its adult form, the tick is about the size of a pinhead or the period at the end of this sentence. It is smaller than the American dog tick, and it lives and mates in the fur of the white-tailed deer. Eventually it falls into the tall grass, where it waits for an unsuspecting dog, rodent, or human to pass by. The tick then attaches to its new host and penetrates into the skin; for the next 24 to 48 hours, it takes a blood meal and swells to the size of a small pea. While sucking the blood, it also defecates its intestinal cells into the wound and, should the tick be infected, spirochetes within the cells will be transmitted. If the tick is observed on the skin, it should be removed with a forceps or tweezers, and the area should be thoroughly cleansed with soap and water and an antiseptic applied.

The diagnosis of Lyme disease is usually based on symptoms, and the physician will often use a pen to mark off the border of the skin

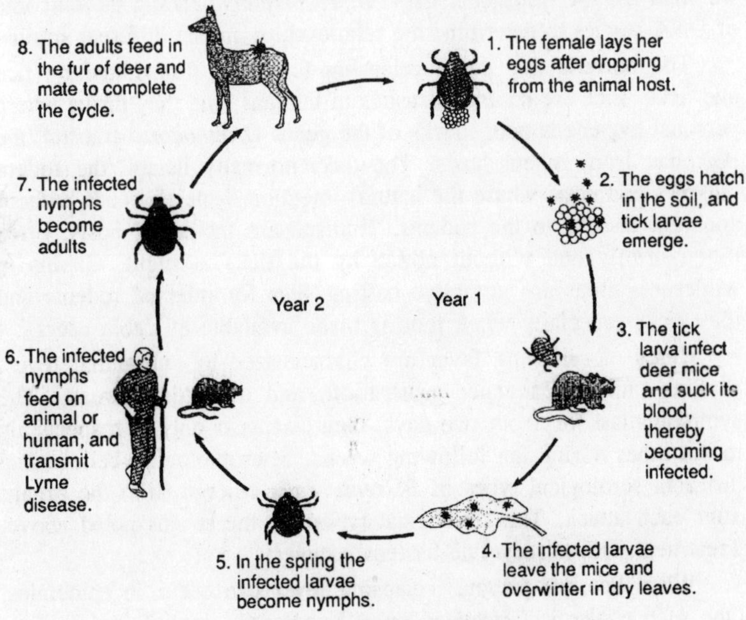


*Fig. 6.1. Life cycle of the tick.*

rash to see if it expands with time. The patient's recent activities are noted (hiking in the woods, living in a tick-infested area, camping), and a blood sample may also be taken for a test for spirochetal antibodies. The test, however, may not be accurate because a number of weeks are required for the body to produce enough antibodies to show up in the test. A newer test designed to detect spirochetal DNA is now being developed.

Congenital transfers of *B. burgdorferi* have been documented, and pregnant women are encouraged to seek early diagnosis and treatment for the disease if they suspect they have been infected. At this writing, a vaccine for dogs has been licensed and is in routine use, but a human vaccine is not yet available. Although Lyme disease is considered a "new" disease, there is evidence that it has existed for some decades even though it has escaped recognition .

**Relapsing Fever**

*Relapsing fever* is caused by a long spirochete currently referred to as *Borrelia recurrentis*. This word "currently" is significant because various names have been given to strains of the spirochete depending on the arthropod vector involved. The CDC, for example, refers to tickborne *Borrelia* species as *B. hermsii* and *B. turicatae*. However, we shall use the traditional name *B. recurrentis* pending the outcome of DNA studies to determine the relationships among different strains.

The borrelia that cause relapsing fever are transmitted by lice and ticks. Lice are natural parasites of humans, and they thrive where personal hygiene is poor. Ticks of the genus *Ornithodorus* transmit the borreliae from rodent hosts. The ticks normally inhabit the rodent burrows and nests where the natural infection cycle proceeds without apparent disease in the rodents. Humans are incidental hosts, often bitten briefly and without notice by the ticks at night. Cabins in wilderness areas are attractive nesting sites for infected rodents and their ticks especially when food is made available by cabin users.

Cases of relapsing fever are characterized by substantial fever, shaking chills, headache, prostration, and drenching sweats. The symptoms last for about two days, then disappear only to reappear up to ten times during the following weeks. Microbiologists believe that different serological types of *B. recurrentis* emerge from the organs after each attack. The serological types are the strains noted above. Treatment with tetracycline hastens recovery.

When lice are present, relapsing fever can occur in epidemics. One such epidemic descended upon Napoleon's lice-infested soldiers during the Russian campaign. Together with serious losses from another louseborne disease typhus, this disease so decimated the French army that the balance tipped in favour of the Russians. Peter Ilyich Tchaikovsky wrote the 1812 Overture to celebrate the great Russian victory, one that might not have been possible without the intervention of microorganisms.

## Rickettsial Arthropodborne Diseases

### Rocky Mountain Spotted Fever

In 1909, Howard Taylor Ricketts, a University of Chicago pathologist, described a new organism in the blood of victims of Rocky Mountain spotted fever and showed that ticks transmit the disease. A year later, he located a similar organism in the blood of animals infected with Mexican typhus, and discovered that fleas were the important vectors in this disease. Unfortunately, in the course of his work, Ricketts fell victim to the disease and died. When later research indicated that Ricketts had described a unique group of microorganisms, the name rickettsiae was coined to honour him. Originally, the rickettsiae were set apart from the bacteria, but microbiologists now consider them to be "small bacteria."

The agent of Rocky Mountain spotted fever is *Rickettsia rickettsii*. This organism is transmitted by ticks, especially those of the genera *Amblyomma* and *Dermacentor*. The hallmarks of Rocky Mountain spotted fever are a high fever lasting for many days and a skin rash reflecting damage to the small blood vessels. The rash begins as pink spots called macules, and progresses to pink-red pimplelike spots known as papules. Where the spots fuse, they form a *maculopapular rash*, which becomes dark red and then fades without evidence of scarring. The rash generally begins on the palms of the hands and soles of the feet and then progressively spreads to the body trunk. Mortality rates of untreated cases are variable, with some outbreaks in Montana recording a rate as high as 75 percent. Treatment with tetracycline or chloramphenicol reduces this rate significantly.

Accurate and rapid diagnosis is essential in Rocky Mountain spotted fever. Evidence of a tick bite, progress of the rash, and the recent activities of the patient (camping, backpacking, and other outdoor activity) are taken into account. Traditionally, a procedure called the *Weil-Felix test* has been employed as well. It is performed by mixing a sample of the patient's serum with the bacterium *Proteus OX19*. The bacteria clump together if the serum contains rickettsial antibodies. Absence of clumping means that the disease must be something else, such as measles or scarlet fever. The test works because rickettsiae

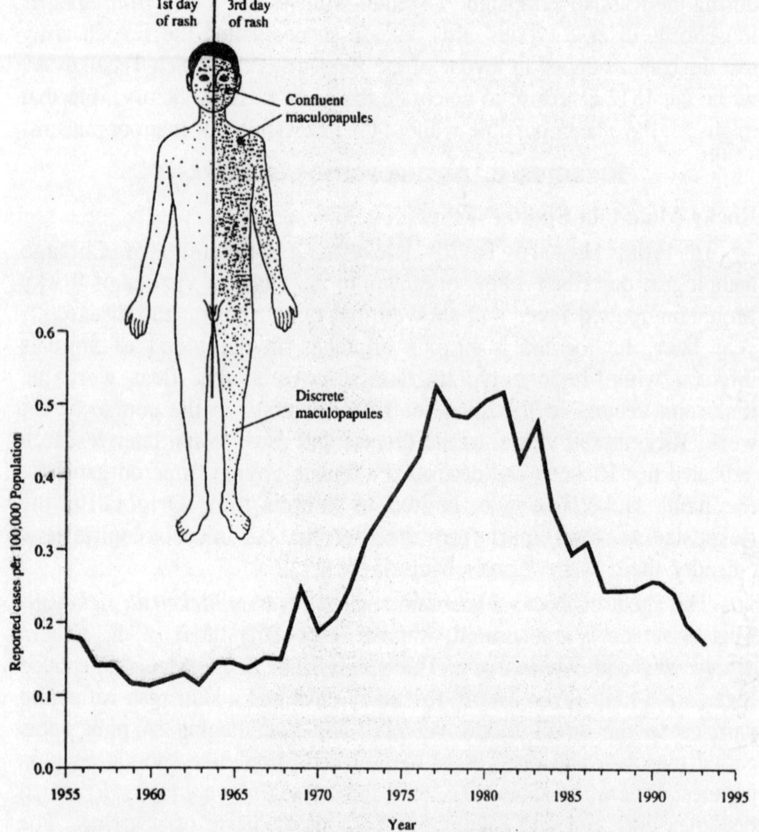


*Fig. 6.2. The graph shows the decline of rocky mountain spotted fever.*

possess antigens also located in *Proteus* cells and, therefore, both organisms stimulate the immune system to produce the same antibodies. However, the test is nonspecific and relatively insensitive, so antibody-detection tests are advised more strongly.

Rocky Mountain spotted fever was first observed in early settlers to the American Northwest. About a thousand cases were reported annually in the United States in the early 1980s, but public education about the disease and improved methods of diagnosis and treatment accompanied a drop to about 600 cases by 1994. Contrary to its name, Rocky Mountain spotted fever is not commonly reported from the West any longer, but it remains a problem in many southeastern and Atlantic coat states. Children are its primary victims because of their contact with ticks.

**Epidemic Typhus**

*Epidemic typhus* (also called *typhus fever*) is among the most notorious of all bacterial diseases. It is considered one of the most prolific killers of humans, and no several occasions it has altered the course of history such as when it helped decimate the Aztec population in the 1500s. Historians report that Napoleon marched into Russia in 1812 with over 200,000 French soldiers but his forces were hit hard by typhus. Hans Zinsser's classic book *Rats*, *Lice* and *History* details events such as these and provides an engaging look at the effects of several infectious diseases on civilization.

Epidemic typhus is caused by *Rickettsia prowazekii*. The bacteria are transmitted among individuals by head and body lice of the genus *Pediculus*. This was first noted by Charles Nicolle, winner of the 1928 Nobel Prize in Physiology or Medicine. Lice are natural parasites of humans. They flourish where sanitation measures are lacking and hygiene is poor. Often these conditions are associated with war, famine, poverty, and a generally poor quality of life.

The characteristic fever and rash of rickettsial disease are particularly evident in epidemic typhus. The rash is maculopapular, but unlike the rash of Rocky Mountain spotted fever, it appears first on the body trunk and then progresses to the extremities. Intense fever, sometimes reaching 104°F, remains for many days as the patient hallucinates and becomes delirious (typhus, like typhoid, takes its name from the Greek word *typhos*, meaning smoke or cloud). Some patients suffer permanent damage to the blood vessels and heart, and over 75 percent of sufferers die in epidemics. Tetracycline and chloramphenicol reduce this figure substantially, and the disease is rare in the United States. The diagnosis of epidemic typhus depends on observation of symptoms, evidence of lice infestation, and a number of serological tests. For definitive diagnosis, samples of patient's serum are injected into guinea pigs, where characteristic changes occur in the reproductive organs. Control of the disease requires destruction of lice populations through good hygiene. Insecticide powders on the clothes and body surface are also of value. During World War II, delousing with DDT was common practice.

**Endemic Typhus**

*Endemic typhus* is a second form of typhus. This disease occurs sporadically in human populations because the flea that transmits it, *Xenopsylla cheopis* is not a natural parasite of humans. However, the disease is prevalent in rodent populations where fleas abound (such as rats and squirrels), and thus it is called *murine typhus* from the Latin *murine*, referring to the mouse. Cats and their fleas are also involved and lice may harbour the bacilli.

The agent of endemic typhus is *Rickettsia typhi*. Both rodent and flea may harbour it for weeks without displaying any effect, but when an infected flea feeds in the human skin, it deposits the organism into the wound. Alternatively, feeds in the human skin, it deposits the organism into the wound. Alternatively, a person may crush the flea while scratching the wound. Endemic typhus is usually characterized by a mild fever, persistent headache, and a maculopapular rash. Often the recovery is spontaneous, without need of drug therapy. However, lice may transport the rickettsiae to other individuals and initiate an epidemic. In the Southwest, endemic typhus is known as *Mexican typhus*, the disease from which Ricketts died. Since first described in 1570, it has also been called *tabardillo* from *tabardo*, meaning coloured cloak, a reference to the mantlelike rash that covers the body. The sharp reduction from 3400 cases in 1941 to about 49 cases in 1987 is mainly due to rodent and flea control.

**Other Rickettsial Diseases**

Several other rickettsial diseases have significance in microbiology because they occur sporadically. These diseases are transmitted among humans by arthropod vectors, and most are characterized by fever and rash. The mortality rates are generally low.

*Scrub typhus* is so-named because it occurs in scrubland where the soil is sandy or marshy and vegetation is poor. *Trombicula*, the mite that transmits the disease, lives in areas such as these. Scrub typhus is also called *tsutsugamushi* from the Japanese words *tsutsuga* for disease, and *mushi* for mite. The causative agent, *Rickettsia tsutsugamushi*, enters the skin during mite infestations and soon causes fever and a rash, together with other typhuslike symptoms. Outbreaks may be significant as evidence by the 7000 U.S. servicemen affected in the Pacific area in World War II. Tetracycline is used in treatment.

*Rickettsialpox* was first recognized in 1946 in an apartment complex in New York City. Investigators traced the disease to mites in the fur of local mice, and named the disease rickettsialpox because the skin rash was similar to that in chickenpox. Rickettsialpox is now considered a benign disease. It is caused by *Rickettsia akari* (*acari* is Greek for mite). Fever and rash are typical symptoms, and fatalities are rare. In many cases the disease runs its course without being recognized. Russia, Korea, and the northeastern United States account for the few cases detected annually. Except for the great influenza pandemic of 1918-1919, *trench fever* was the most widespread disease encountered during World War I. An estimated million soldiers are thought to have been infected. The disease is caused by *Rochalimaea quintana*, one of the few rickettsiae cultivated outside of living cells. Transmission takes place by head and body lice, and symptoms include a maculopapular rash and a fever occurring at irregular intervals. Trench fever is prevalent where lice abound, and trench warfare provided a near-ideal setting. Mortality rates are generally low. The disease is sometimes called *His-Werner disease* for the two investigators who studied it early in this century.

In various regions of the world, *Rickettsia conorii* caused a series of *tickborne fevers*. These diseases are known as boutonneuse fever, Marseilles fever, Nigerian typhus, and South African tick bite fever. All appear to be mild versions of Rocky Mountain spotted fever and all are accompanied by a fever and rash. In addition, a black spot (*la tache noire*) develops where the tick has fed in the skin. Wild rodents harbour the disease in nature, and outbreaks occurs sporadically when they make contact with human populations.

In textbooks from the early part of this century an illness called *Brill-Zinsser disease* was described as a mil form of typhus. The high fever was a distinguishing characteristic, and it was recognized that if a patient was infested with lice, an epidemic of typhus might ensue. Today microbiologists believe that Brill-Zinsser disease is a recurrence of an earlier case of typhus in which *Rickettsia prowazekii* lay dormant in the patient for many years.

*Ehrlichiosis*, the final disease we shall consider, was first described in humans in 1986. Formerly believed confined to dogs ehrlichiosis has been recognized in two forms in the United States: *human monocytic ehrlichiosis* (HME) caused by the rickettsia *Ehrlichia chaffeensis* (because the first case was observed at Fort Chaffee, Arkansas); and *human granulocytic ehrlichiosis* (HGE) due to *E. equi*, also a rickettsia. Patients suffer headache, malaise, and fever, but no rash or other distinctive physical findings. There is also a marked reduction of the platelet (thrombocyte) count, a condition known as thrombocytopenia. HME affects the monocytes ("monocytic"), while HGE affects the neutrophils ("granulocytic"). The agents of both diseases are transmitted by ticks, and tetracycline antibiotics are used in therapy. A notable outbreak of HGE affected 68 patients in the New York City area in 1994.

# 7

# Physical Control

By the best estimates of health officials the problem began during the first week of July in 1988. One day that week a person unintentionally contaminated the water while swimming in an indoor pool at a school in Los Angeles, California. The same week the pool was used by a water polo team for its match against a local opponent. Shortly thereafter, a class of SCUBA divers used the pool to perfect their skills. A group of elementary school children also enjoyed the pool during a camp field trip.

Beginning in early August, doctors began receiving reports of watery diarrhea, abdominal cramps, and fever. Two persons from the SCUBA class were hospitalized. Eleven patients were tested for stool microorganisms and in seven cases, the protozoan *Cryptosporidium* was located. Indeed, the attack rate was highest for those who were exposed to the water the longest. By the second week of August, 44 persons were affected from a total of 60 swimmers during that fateful July.

When CDC investigators arrived at the pool they inquired about the water purification methods and were told that three filters were used to sterilize the water and chlorine was added to maintain a low bacterial count. Their inspection revealed that the chlorination was adequate, but the filtration left much to be desired. One of the three filters was inoperative, and the flow rate of water through the remaining two filters was only 70 percent of the expected rate. The water filtration had not worked.

The episode in Los Angeles illustrates how microorganisms can spread when proper methods of water purification are lacking. In many cases, these methods are designed to achieve sterilization. *Sterilization* implies the destruction or removal of all life forms. It is an absolute term that cannot be qualified. Thus one cannot assume "partial sterilization" or "incomplete sterilization," but must consider a material "contaminated" until sterilized. It should be pointed out, however, that material may remain harmful even though it is sterile. For example, a solution of bacterial toxin may contain no living forms but still cause physiological damage in the body.

Although chemicals may sometimes be used to sterilize objects, the principal methods for achieving sterilization employ physical agents such as heat, as well as radiations, and filtration. These methods are not products of the modern ear. They were used by Pasteur, Koch, and microbiologists of a century ago to prevent contamination of their materials and to ensure the accuracy of their work.

Microbiologists recognize that bacterial *spores* are the most resistant forms of life. Anthrax spores, for example have been found viable after 60 years on dry silk threads. The spores are not metabolically inert. Rather, they carry on life processes at ninimum rates and possess the necessary enzymes to transform from a dormant state to actively metabolizing vegetative cells. The extraordinary survival of spores is often attributed to their low water content, which yields a heat-resistant gel-like spore core. Another possible reason is the presence of *dipicolinic acid*, an organic compound that encourages heat resistance by linking to the spore proteins. Destruction of the bacterial spore is the principal aim of sterilization methods, especially those involving heat.

## Physical Control with Heat

*The Citadel* is a novel by A.J. Cronin that follows the life of a young British physician, beginning in the 1920s. Early in the story the physician, Andrew Manson, begins his practice in a small coal-mining town in Wales. Almost immediately, he encounters an epidemic of *typhoid fever*. When his first patient dies of the disease, Manson becomes terribly distraught. However, he realizes that the epidemic can be halted, and in the next scene, he is tossing all of the patient's bedsheets, clothing, and personal effects into a huge bonfire.

The killing effect of *heat* on microorganisms has long been known. Heat is fast, reliable, and relatively inexpensive, and it does not introduce chemicals to a substance, as disinfectants sometimes do. Above maximum growth temperatures, biochemical changes in the cell's organic molecules result in its death. These changes arise from alterations in enzyme molecules or chemical breakdowns of structural molecules, especially in the cell membranes. Heat also drives off water, and since all organisms depend on water, this loss may be lethal.

The killing rate of heat may be expressed as a function of time and temperature. For example, *tubercle bacilli* are destroyed in 30 minutes at 58°C, but in only 2 minutes at 65°C, and in a few seconds at 72°C. Each microbial species has a *thermal death time*, the time necessary for killing it at a given temperature. Each species also has a *thermal death point*, the temperature at which it dies in a given time. These measurements are particularly important in the food industry, where heat is used for preservation.

In determining the time and temperature for microbial destruction with heat, certain factors bear consideration. One factor is the type of organism to be killed. For example, if materials are to be sterilized, the physical method must be directed at bacterial spores. Milk, however, need not be sterile for consumption, and heat is therefore aimed at the most resistant vegetative cells of pathogens.

Another factor is the type of material to be treated. Powder is subjected to dry heat rather than moist heat, because moist heat will

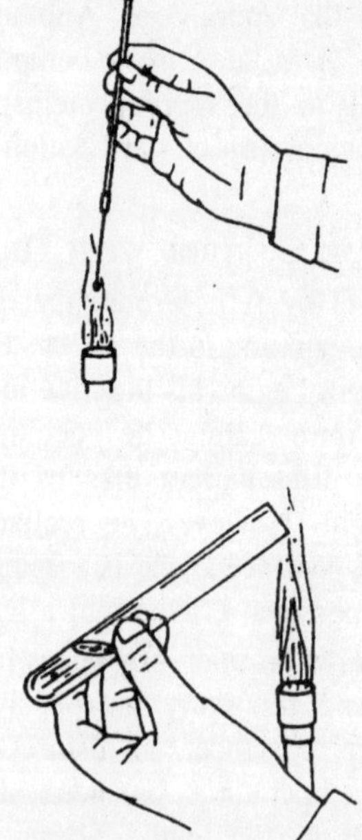

*Fig. 7.1. Laboratory use of the Bunsen burner.*

leave it soggy. Saline solutions, by contrast, can be sterilized with moist heat but are not easily treated with dry heat. Other factors are the presence of organic matter and the acidic or basic nature of the material. Organic matter may prevent heat from reaching microorganisms, while acidity or alkalinity may encourage the lethal action of heat.

**Direct Flame**

Perhaps the most rapid sterilization method is the *direct flame* method used in the process of incineration. The flame of the Bunsen burner is employed to sterilize the bacteriological loop before removing a sample from a culture tube and after preparing a smear. Flaming the tip of the tube also destroys organisms that happen to contact the tip, while burning away lint and dust.

In general, objects must be disposable if a flame is used for sterilization. Disposable hospital gowns and certain plastic apparatus are examples of materials that may be incinerated. In past centuries, the bodies of disease victims were burned to prevent spread of the pestilence. It is still common practice to incinerate the carcasses of cattle that have died of *anthrax* and to put the contaminated field to the torch because anthrax spores cannot adequately be destroyed by other means. British law even stipulates that anthrax-contaminated animals may not be autopsied before burning.

**Hot-Air Oven**

The *hot-air oven* utilizes radiating dry heat for sterilization. This type of energy does not penetrate materials easily and thus, long periods of exposure to high temperatures are necessary. For example, at a temperature of 160°C (320°F), a period of two hours is required for the destruction of bacterial spores. Higher temperatures are not recommended because the wrapping paper used for equipment tends to char at 180°C. The hot-air method is useful for sterilizing dry powders and water-free oily substances, as well as many types of glassware such as pipettes, flasks, and syringes. Dry heat does not corrode sharp instruments as steam often does, nor it erode the ground glass surfaces of nondisposable syringes.

The effect of *dry heat* on microorganisms is equivalent to that of baking. The heat changes microbial proteins by oxidation reactions and creates an arid internal environment, thereby burning microorganisms slowly. It is essential that organic matter such as oil or grease film be removed from the materials, because organic matter insulates against dry heat. Moreover, the time required for heat to reach sterilizing temperatures varies among materials. Thus this factor must be considered in determining the total exposure time.

**Boiling Water**

Immersion in *boiling water* is the first of several moist-heat methods that we shall consider. *Moist heat* penetrates materials much more rapidly than dry heat because water molecules conduct heat better than air. Lower temperatures and less time of exposure are therefore required than for dry heat.

Moist heat kills microorganisms by denaturing their proteins. *Denaturation* involves changes in the chemical or physical properties

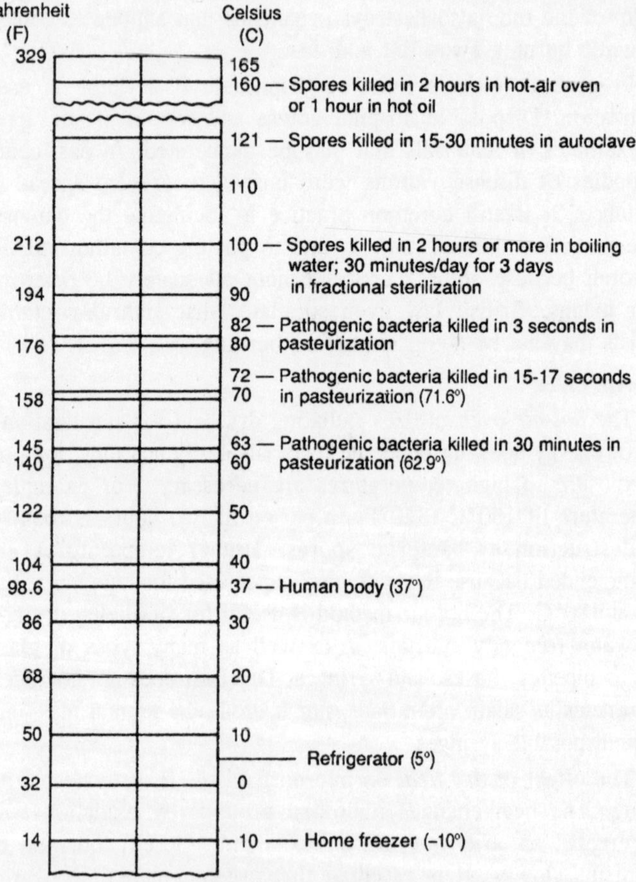


*Fig. 7.2. Various temperature environments used in the physical control of microorganisms.*

of proteins. It includes structural alterations due to destruction of the chemical bonds holding proteins in a three-dimensional form. As proteins revert to a two-dimensional structure, they coagulate (denature) and become nonfunctional. Egg protein undergoes a similar transformation when it is boiled. The chemical structure of proteins helpful to your understanding of this process. The coagulation and denaturing of proteins require less energy than oxidation, and therefore, less heat need be applied.

Boiling water is not considered a sterilizing agent because destruction of bacterial spores and inactivation of viruses cannot always be assured. Under ordinary circumstances, with microorganisms at concentrations of less than one million per milliliter, most species of microorganisms can be killed within 10 minutes. Indeed, a few seconds may be all that is necessary. However, fungal spores, protozoal cysts, and large concentrations of hepatitis A viruses require up to 30 minutes' exposure. Bacterial spores often require two hours or more. Because inadequate information exists on the heat tolerance of many microorganisms, boiling water is not reliable for sterilization purposes.

If it is imperative that boiling water be used to destroy microorganisms, materials must be thoroughly cleaned to remove traces of organic matter such as blood or feces. The minimum exposure period should be 30 minutes, except at high altitudes, where it should be increased to compensate for the lower boiling point of water. All materials should be well covered. Washing soda may be added at a 2 percent concentration to increase the efficiency of the process.

**Autoclave**

Moist heat in the form of pressurized steam is regarded as the most dependable method for the destruction of all forms of life, including bacterial spores. This method is incorporated into a device called the *autoclave*. Over a hundred years ago, French and German microbiologists developed the autoclave as an essential component of their laboratories.

A basic tenet of chemistry is that when the pressure of a gas increases, the temperature of the gas increases proportionally. Because steam is a gas, increasing its pressure in a closed system increases its temperature. As the water molecules in steam become more energized, their penetration increases substantially. This principle is used to reduce cooking time in the home pressure cooker and to reduce sterilizing time in the autoclave. It is important to note that the sterilizing agent is the moist heat, not the pressure.

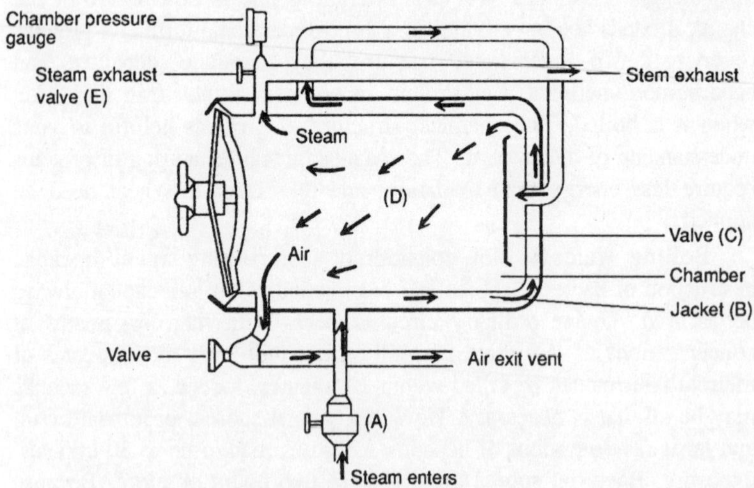


*Fig. 7.3. Operation of the autoclave.*

Most autoclaves contain a *sterilizing chamber* into which articles are placed and a steam jacket where steam is maintained. As steam flows from the steam jacket into the sterilizing chamber, cool air is forced out and a special valve increases the pressure to 15 ponds/ square inch (lb/in$^2$) above normal atmospheric pressure. The temperature rises to 121.5°C and the superheated water molecules rapidly conduct heat into microorganism. The time for destruction of the most resistant bacterial spore is now reduced to about 15 minutes. For denser objects, up to 30 minutes of exposure may be required.

The autoclave is used to control microorganisms in both hospitals and laboratories. It is employed for blankets, bedding, utensils, instruments, intravenous solutions, and a broad variety of other objects. The laboratory technician uses it to sterilize bacteriological media and destroy pathogenic cultures. The autoclave is equally valuable for glassware and metalware, and is among the first instruments ordered when a microbiology laboratory is established.

The autoclave also has certain *limitations*. For example, some plasticware metals in the high heat, and sharp instruments often become dull. Moreover, many chemicals break down during the sterilization process, and oily substances cannot be treated since they do not mix with water. To gauge the success of sterilization, a strip containing spores of a *Bacillus* species is included with the objects treated. At the conclusion of the cycle, the strip is placed in nutrient broth medium and incubated. If the sterilization process has been successful, no growth will occur, but growth indicates failure. In recent years a new form of autoclave, called the *prevacuum autoclave*, has been developed for sterilization procedures. This machine draws air out of the sterilizing chamber at the beginning of the cycle. Saturated steam is then used at a temperature of 132°C to 134°C at a pressure of 28 to 30 lb/in$^2$. the time for sterilization is now reduced to as little as 4 minutes. A vacuum pump operates at the end of the cycle to remove the stem and dry the load. The major advantages of the prevacuum autoclave are the minimal exposure time for sterilization and the reduced time to complete the çycle.

**Fractional Sterilization**

In the years before development of the autoclave, liquids and other objects were sterilized by exposure to free-flowing steam at 100°C for 30 minutes on each of three successive days. The method was called *fractional sterilization* because a fraction was accomplished on each day. It was also called *tyndallization* after its developer, John Tyndall and *intermittent sterilization* because it was a stop-and-start operation.

Sterilization by the fractional method is achieved by an interesting series of events. During the first day's exposure, steam kills virtually all organisms except bacterial spores, and its stimulates spores to germinate to vegetative cells. During overnight incubation the cells multiply and are killed on the second day. Again, the material is cooled and the few remaining spores germinate, only to be killed on the third day. Although the method usually results in sterilization, occasions arise when several spores fail to germinate. The method also requires that spores be in a suitable medium for germination, such as a broth. Fractional sterilization has assumed importance in modern microbiology with the development of high-technology instrumentation and new chemical substances. Often, these materials cannot be sterilized at autoclave temperatures, or by long periods of boiling or baking, or with chemicals. An instrument that generates free-flowing steam, such as the *Arnold sterilizer*, is used in these instances.

**Pasteurization**

*Pasteurization* is not the same as sterilization. Its purpose is to reduce the bacterial population of a liquid such as milk and to destroy organisms that may cause spoilage and human disease. Spores are not affected by pasteurization. One method for milk pasteurization, called the *holding method*, involves heating at 62.9°C for 30 minutes. Although thermophilic bacteria thrive at this temperature, they are of little consequence because they cannot grow at body temperature. For decades, pasteurization has been aimed at destroying *Mycobacterium tuberculosis*, long considered the most heat-resistant bacterium. More recently, however, attention has shifted to destruction of *Coxiella burnetii*, the agent of Q fever, because this organism has a higher resistance to heat. Since both organisms are eliminated by pasteurization, dairy microbiologists assume that other pathogenic bacteria are also destroyed. Two other methods of pasteurization are the *flash pasteurization* method at 71.6°C for 15 seconds, and the *ultra-pasteurization* method at 82°C for 3 seconds.

### Hot Oil

Some dentists and physicians use *hot oil* at 160°C for the sterilization of instruments. A time period of 1 hour is usually recommended. Hot oil does not rust metals, and minimal corrosion takes place. However, once sterilization is complete, the instruments must be cleaned and dried for storage, and this step may reintroduce contamination. Silicone is sometimes used as an alternative to oil.

## Physical Control by Other Methods

### Filtration

Heat is a valuable physical agent for controlling microorganisms but sometimes it is impractical to use. For example, no one would suggest removing the microbial population from a tabletop by using a Bunsen burner, nor can heat-sensitive solutions be subjected to an autoclave. In instances such as these and numerous others, a heat-free method must be used. This section describe some examples.

Filters came into prominent use in microbiology as interest in viruses grew in the 1890s. Previous to that time, filters had been utilized to trap airborne organisms and sterilize bacteriological media, but now they become essential for separating viruses from other microorganisms. Among the early pioneers of filter technology was Charles Chamberland, an associate of Pasteur. His porcelain filter was important to early virus research. Another pioneer was *Julius Petri* (inventor of the Petri dish), who developed a sand filter to separate bacteria from the air. The *filter* is a mechanical device for removing microorganisms from a solution. As fluid passes through the filter, organisms are trapped in the pores of the filtering material. The solution that drips into the receiving container is decontaminated or, in some cases, sterilized. Filters are used to purify such things as intravenous solutions, bacteriological media, toxoids, many pharmaceuticals, and beverages.

Several types of filters are available for use in the microbiology laboratory. *Inorganic filters* are typified by the Seitz filter, which consists of a pad of porcelain or ground glass mounted in a filter flask. *Organic filters* are advantageous because the organic molecules of the filter attract organic components in microorganisms. One example, the Berkefeld filter, utilizes a substance called *diatomaceous earth*. This material contains the remains of marine algae known as diatoms. *Diatoms* are unicellular algae that abound in the oceans and provide important foundations for the world's food chains. Their remains accumulate on the shoreline and are gathered for use in swimming pool and aquarium filters, as well as for microbiological filters used in laboratories.

The *membrane filter* is a third type of filter that has received broad acceptance. It consists of a pad of organic compounds such as cellulose acetate or polycarbonate, mounted in a holding device. This filter is particularly valuable because bacteria multiply and form colonies on the filter pad when the pad is placed on a plate of culture medium. Microbiologists can then count the colonies to determine the number of bacteria originally present. For example, if a 100 ml sample of liquid were filtered and 59 colonies appeared on the pad after incubation, it could be assumed that 59 bacteria were in the sample.

Air can also be filtered to remove microorganisms. The filter generally used is a *high-efficiency particulate air (HEPA) filter*, which can remove over 99 percent of all particles, including microorganisms having a diameter of over 0.3 micrometer. The air entering surgical units and specialized treatment facilities such as burn units is filtered to exclude microorganisms. In some hospital wards, such as for respiratory diseases, and in certain pharmaceutical filling rooms, the air is recirculated through HEPA filters to ensure its purity.

**Ultraviolet Light**

*Visible light* is a type of radiant energy detected by the sensitive cells of the eye. The wavelength of this energy is between 400 and 800 nanometers (nm). Other types of radiations have wavelengths longer or shorter than that of visible light and therefore, they cannot be detected by the human eye.

One type of radiant energy, *ultraviolet light*, is useful for controlling microorganisms. Ultraviolet light has a wavelength between 100 to 400 nm, with the energy at about 265 nm most destructive to bacteria. When microorganisms are subjected to ultraviolet light, cellular DNA absorbs the energy, and adjacent *thymine molecules* link together. Linked thymine molecules are unable to position adenine on messenger RNA molecules during the process of protein synthesis. Moreover, replication of the chromosome in binary fission is impaired. The damaged organism can no longer produce critical proteins or reproduce, and it dies quickly.

Ultraviolet light effectively reduces the microbial population where direct exposure takes place. It is used to limit airborne or surface contamination in a hospital room, morgue, pharmacy, toilet facility, or food service operation. It is noteworthy that ultraviolet light from the sun may be an important factor in controlling microorganisms in the air and upper layers of the soil, but it may not be effective against all bacterial spores. Ultraviolet light does not penetrate liquids or solids, and it may cause damage in human skin cells.

**Other Types of Radiation**

The spectrum of energies includes two other forms of radiation useful for destroying microorganisms. These are *X rays* and *gamma rays*. Both have wavelengths shorter than the wavelength of ultraviolet light. As X rays and gamma rays pass through microbial molecules, they force electrons out of their shells, thereby creating ions. For this reason, the radiations are called *ionizing radiations*. The ions quickly combine with and destroy proteins and nucleic acids such as DNA, causing death. Gram-positive bacteria are more sensitive to ionizing radiations than Gram-negative bacteria. Ionizing radiations are currently used to sterilize such heat-sensitive pharmaceuticals as vitamins, hormones, and antibiotics as well as certain plastics and suture materials.

Another form of energy, the *microwave*, has a wavelength longer than that of ultraviolet light. In a microwave oven, microwaves are absorbed by water molecules. The molecules are set into high-speed motion, and the heat of friction is transferred to foods, which become hot rapidly. Other than the heat generated, there is no specific activity against microorganisms. A final form of radiation that we shall consider is light energy. When concentrated by sophisticated devices, light energy forms a *laser beam*. The term "laser" stands for light amplification by stimulated emission of radiation. Recent experiments indicate that laser beams can be used to sterilize instruments and the air in operating rooms, as well as for a wound surface. Microorganisms are destroyed in a fraction of a second, but the laser beam must reach all parts of the material to effect sterilization.

**Ultrasonic Vibrations**

*Ultrasonic vibrations* are high-frequency sound waves beyond the range of the human ear. When directed against environmental surfaces,

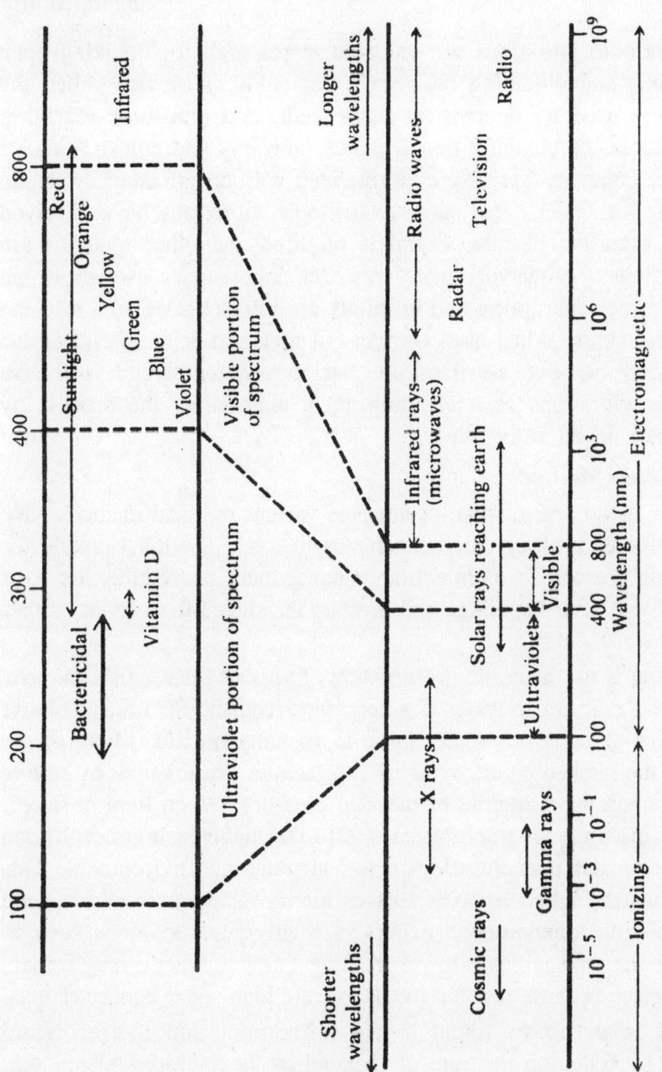


*Fig. 7.4. The ionizing and electromagnetic spectrum of energies.*

they have little value because air particles deflect and dispersed the vibrations. However, when propagated in fluids, ultrasonic vibrations cause the formation of microscopic bubbles, or cavities, and the water appears to boil. Some observers call this "cold boiling". The cavities rapidly collapse, and send out shock waves. Microorganisms in the fluid are quickly disintegrated by the external pressures. The formation and implosion of the cavities is known as *cavitation*.

Ultrasonic vibrations are valuable in research for breaking open tissue cells and obtaining their parts for study. A device called the *cavitron* is used by dentists to clean teeth, and ultrasonic machines are available for cleaning dental plates, jewelry, and coins. A major appliance company has also experimented with an ultrasonic washing machine. As a sterilizing agent, ultrasonic vibrations have received minimal attention because liquid is required and other methods are more efficient. However, many research laboratories use ultrasonic probes for cell disruption and hospitals use ultrasonic devices to clean their instruments. When used with an effective germicide, an ultrasonic device may achieve sterilization, but the current trend is to use ultrasonic vibrations as a cleaning agent and follow the process by sterilization in an autoclave.

**Preservation Methods**

Over the course of many centuries, various physical methods have evolved for controlling microorganisms in food. Though valuable for preventing the spread of infectious agents, these procedures are used principally to retard spoilage and prolong the shelf life of foods, rather than for sterilization.

*Drying* is useful in the preservation of various meats, fish, cereals, and other foods. Since water is a necessary requisite for life, it follows that where there is no water, there is virtually no life. Many of the foods in the kitchen pantry typify this principle. Preservation by *salting* is based upon the principle of osmotic pressure. When food is salted, water diffuses out of microorganisms to the higher salt concentration and lower water concentration in the surrounding environment. This flow of water, called *osmosis*, leaves microorganisms to shrived and die. The same phenomenon occurs in highly sugared foods such as syrups, jams, and jellies. However, fungal contamination may remain at the surface because aerobic molds tolerate high sugar concentrations.

*Low temperatures* found in the refrigerator and freezer retard spoilage by reducing the rate of metabolism in microorganisms and, consequently, reducing their rate of growth. Spoilage is not totally eliminated in cold foods, however, and many microorganisms remain alive, even at freezer temperatures. These organisms multiply rapidly when food thaws, which is why prompt cooking is recommended. Note in these examples that there are significant differences between killing microorganisms, holding them in check, and reducing their numbers. The preservation methods are described as bacteriostatic because they prevent the further multiplication of bacteria.

# 8

# CHEMICAL CONTROL

Before the 1900s, hospitals rarely had running water, and what water they had was usually contaminated. Garbage, human waste, and other hospital refuse were usually dumped into a pit in the courtyard; surgeons wiped their hands and instruments on their hospital jackets and trousers; bedclothes were rarely changed; and infection was rampant. Up to one-third of women giving birth died of puerperal fever, a blood disease often due to a species of *Streptococcus*.

As late as the mid-1800s, only a few visionaries recognized the relationship between filth and disease. One such person was the American physician, poet, and jurist, Oliver Wendell Holmes, author of an 1843 paper on the contagious nature of puerperal fever. Another was the Hungarian physician *Ignaz Semmelweis*.

In 1847, Semmelweis was working at a Vienna hospital's obstetrics clinic when he made a remarkable observation: unusually high incidence of puerperal fever occurred in maternity wards tended by physicians fresh from dissecting cadavers. However, in the ward tended by midwives, the incidence was much lower. Semmelweis reasoned that disease was transmitted by infected hands, and he ordered all the attendants in his ward to wash their hands in chlorine water before ministering to patients. Soon the death rate among his patients fell significantly.

Semmelweis summarized his findings and presented his evidence for the transmission of puerperal fever to hospital administrators. However, they rejected his conclusion that doctors were spreading disease because it cast doctors in a negative light. The debate became vehement, and Semmelweis insisted that rejecting his methods was tantamount to committing murder. Shortly thereafter, an anguished and frustrated Semmelweis left Austria for his native Hungary, where he died in 1865.

Ironically, the death of Semmelweis paralleled Pasteur's landmark experiments in microbiology. As the germ theory of disease gained a foothold, the realization dawned that infectious microorganisms could indeed be transmitted by clothing, utensils, instruments, and other objects. To interrupt the spread of organisms, doctors began using chemical antiseptics and disinfectants, and before long they witnessed a substantial decline in the incidence of disease.

In this chapter we shall examine a variety of chemical methods used for controlling the spread of microorganisms. Our study begins by outlining some general principles and terminology of disinfection practices, and then proceeds to a discussion of the spectrum of antiseptics and disinfectants. Whether for wounds, swimming pools, industrial machinery, or pharmaceutical products, antiseptics and disinfectants are fundamental to public health practices that ensure continued good health.

## General Principles of Chemical Control

The notions of sanitation and disinfection are not unique to the modern era. *The Bible* refers often to cleanliness and prescribes certain dietary laws to prevent consumption of wheat was believed to be contaminated food. Egyptians used resins and aromatics for embalming even before they had a written language, and ancient peoples burned sulfur for deodorizing and sanitary purposes. Over the centuries, necessity demanded chemicals for food preservation, and spices were used as preservatives as well as masks for foul odours. Indeed, Marco Polo's trips to the Orient for new spices were made out of necessity as well as adventure.

Medicinal chemicals came into widespread use in the 1800s. As early as 1830, for example, the *U.S. Pharmacopoeia* listed tincture of iodine as a valuable antiseptic, and soldiers in the Civil War used it in plentiful amounts. In the first decades of that century, people found copper sulfate useful for preventing fungal disease in plants, but unfortunately, this chemical was not well-known during the great potato blight of Ireland. Mercury was sometimes used for treating syphilis, as first suggested by Arabian physicians centuries before. Moviegoers have probably noted that American cowboys practiced disinfection by pouring whiskey onto gunshot wounds between drinks.

In the 1860s, a physician at the University of Glasgow named *Joseph Lister* established the principles of aseptic surgery. Pasteur's references to airborne microorganisms convinced Lister that microorganisms were the cause of wound infections. Lister experimented with several chemicals to kill microorganisms and finally settled on carbolic acid (phenol). He applied the chemical to wounds and instruments and sprayed it in the air near the operating table. Lister achieved brilliant success and was accorded many accolades as a pioneer microbiologist of his time, although he advocated the same principles recommended by Semmelweis years before.

**Terminology of Disinfection**

The chemical control of microorganisms extends into such diverse areas as the hospital environment, food processing plant, and everyday household. These broad fields have yielded an equally broad terminology that should be explored before we undertake a discussion of individual chemical agents.

The physical agents for controlling microorganisms are generally intended to achieve *sterilization*. This implies the destruction of all forms of life, especially bacterial spores. Chemical agents, by contrast, rarely achieve sterilization. Instead, they are expected only to destroy the pathogenic organisms on or in an object. The process of destroying pathogens is called *disinfection*; the object is said to be disinfected. If the object is lifeless, such as a table top, the chemical agent is known as a *disinfectant*. However, if the object is living, such as a tissue of the human body, the chemical is an *antiseptic*. It is important to note that even though a particular chemical may be used in a disinfectant as well as an antiseptic (iodine, for example), the precise formulations are so different that the ability to kill microorganisms differs substantially in the two products.

Antiseptics and disinfectants are usually bactericidal, but occasionally they may be bacteriostatic. A *bactericidal agent* kills microorganisms, while a *bacteriostatic agent* temporarily prevents their further multiplication without necessarily killing them. For example, a bactericidal chemical may inactivate the enzymes of an organism and interfere with its metabolism so that it dies. A bacteriostatic chemical, by contrast, disrupts a minor chemical reaction and slows the metabolism, resulting in a longer time between divisions. Although a delicate difference sometimes exists between the bactericidal and bacteriostatic nature of a chemical, the terms indicate effectiveness of the chemical agent in a particular situation.

(a) Antiseptic

(b) Disinfectant

*Fig. 8.1. Sample uses of antiseptics and disinfectants.*

The word *sepsis* is derived from the Greek *seps*, meaning putrid. It refers to contamination of an object by microorganisms and forms the steams for "septicemia," meaning microbial infection of the blood, and "antiseptic," which translates to "against infection." It is also the origin of the term *aseptic*, meaning "free of contaminating microorganisms."

Other expressions associated with chemical control are sanitize and degerm. To *sanitize* an object is to reduce the microbial population to a safe level as determined by local public health standards. For example, in dairy and food processing plants, the equipment is usually sanitized. To *degerm* something is merely to remove organisms from a surface. Washing with soap and water degerms the skin surface but has little effect on microorganisms deep in the skin pores.

A final group of terms that bears mention is the -cidal agents. These include the fungicidal agents, which kill fungi; the virucidal agents, for viruses; the sporicidal agents, for bacterial spores; and the germicidal agents, for various types of microorganisms.

## Selection of Antiseptics and Disinfectants

To be useful as an antiseptic or disinfectant, a chemical agent must have certain properties, some of which are more desirable than others. The first prerequisite is that it must be able to kill microorganisms. It should also be nontoxic to animals or humans, especially if it is used as an antiseptic. It should be soluble in water and have a substantial shell life during which its activity is retained. The agent should be useful in very diluted form and perform its job in a relatively short time. Both factors tend to reduce toxic side effects.

Other characteristics will also contribute to the value of a chemical agent: it should not separate on standing; it should penetrate well; and it should not corrode instruments. The chemical will have a distinct advantage if it does not combine with organic matter such as blood or feces, because microorganisms are often found in this type of material. Of course, the chemical should be easy to obtain and relatively inexpensive. Since disinfection is essentially a chemical process, the *parameters of chemistry* should be considered when selecting an antiseptic or disinfectant. For example, the temperature at which the disinfection is to take place may be important because a chemical reaction occurring at 37°C (body temperature) may not occur at 25°C (room temperature). Also, a particular chemical may be effective at a certain pH but not another. Moreover, the chemical reaction may be very rapid with one agent and slower with another. Thus if long-term disinfection is desired, the second agent may be preferable.

Two other considerations are the type of microorganism to be eliminated and the surface treated. For instance, the removal of bacterial spores requires more vigorous treatment than the removal of vegetative cells. Also a chemical applied to a laboratory bench is considerably different from one used on a wound. It is therefore imperative to distinguish the antiseptic or disinfectant nature of a chemical before proceeding with use. Indeed, chemical agents formulated as disinfectants are regulated and registered by the federal Environmental Protection Agency, while chemicals formulated as antiseptics are regulated by the federal Food and Drug Administration.

## Evaluation of Antiseptics and Disinfectants

At the current time, the United States Environmental Protection Agency lists over 8000 disinfectants for hospital use and thousands more for general use. Evaluating these chemical agents is a tedious process because of the broad diversity of conditions under which they are used.

A standard of effectiveness used for the chemical agents is the *phenol coefficient* (PC). This is a number that indicates the disinfecting ability of an antiseptic or disinfectant in comparison to phenol under identical conditions. A PC higher than 1 indicates that the chemical is more effective than phenol; a number less than 1 indicates poorer disinfecting ability than phenol. For example, antiseptic A may have a PC of 78.5, while antiseptic B has a PC of 0.28. These numbers are used relative to each other rather than to phenol because phenol is allergenic and irritating to tissues and thus is rarely used.

The phenol coefficient is determined by a laboratory procedure in which dilutions of phenol and the test chemical are mixed with standardized bacteria such as *staphylococcus aureus* and *Salmonella typhi* or other species. The laboratory technician then determines which dilutions have killed the organisms after a 10-minute exposure but not after a 5-minute exposure. The test has many *drawbacks*, especially since it is performed in the laboratory rather than in a real-life situation. Nor does it take into account such factors as tissue toxicity, activity in the presence of organic matter, or variations of temperature.

A more practical way of determining the value of a chemical agent is by an *in-use test*. For example, swab samples of a floor may be taken before and after application of a disinfectant to determine the level of kill. Another method is to dry standardized cultures of bacteria within small stainless steel cylinders and then expose the cylinders to the test chemical. After an established time, the bacteria are assayed for survival to the chemical. These methods of standardization have value under certain circumstances, but it is conceivable that a universal test may never be developed in view of the huge variety of chemical agents available and the numerous conditions under which they are employed.

## Important Chemical Agents

The chemical agents currently in use for controlling microorganisms range from the very simple substances such as halogen ions to the very complex compounds typified by the detergents. Many of these agents in nature have been employed for generation, while other represent the latest developments. In this section we shall survey several groups of chemical agents and indicate how they are best applied in the chemical control of microorganisms.

### Halogens

The *halogens* are a group of highly reactive elements whose atoms have seven electrons in the outer shell. Two halogens, chlorine and iodine, are commonly used for disinfection. *Chlorine* is available in a gaseous form and as organic and inorganic compounds. It is widely used in municipal water supplies, where it keeps bacterial populations at low levels. Chlorine combines readily with numerous ions in water and therefore, enough chlorine must be added to ensure that a residue remains for antibacterial activity. In municipal water, the residue of chlorine is usually about 0.2 to 1.0 parts per million (ppm) of free chlorine. One ppm is equivalent to 0.0001 percent, an extremely small amount.

Chlorine is also available as *sodium hypochlorite* (NaOCl) or as *calcium hypochlorite* ($Ca(OCl)_2$). The latter, also known as chlorinated lime, was used be Semmelweis in his studies in Vienna. Hypochlorite compounds release free chlorine in solution. They are typified by the 0.5 percent sodium hypochlorite solution of H.D. Dakin used extensively for wounds sustained in World Wars I and II. Dakin's solution remains popular in Europe, where it is used to treat athlete's foot.

Sodium hypochlorite is used as a bleaching agent in the textile industry; commercially available *bleach* contains about 5 percent of this compound. To disinfect *clear water*, the Centers for Disease Control and Prevention recommends a half-teaspoon of household chlorine bleach in 2 gallons of water, with 30 minutes contact time before consumption. Hypochlorites are also used in very dilute solutions for disinfecting swimming pools and sanitizing factory equipment.

The *chloramines* such as chloramine-T are organic compounds that contain chlorine. These compounds release free chlorine more slowly than hypochlorite solutions and are more stable. They are valuable for general wound antisepsis and root canal therapy.

Chlorine is effective against a broad variety of organisms including most Gram-positive and Gram-negative bacteria, and many viruses, fungi, and protozoa. However, it is not sporicidal. In microorganisms, the halogen is believed to cause the release of atomic oxygen, which then combines with and inactivates certain cytoplasmic proteins such as enzymes. Another theory is that chlorine changes the structure of cell membranes, thus leading to leakage.

The *iodine* atom is slightly larger than the chlorine atom and is more reactive and more germicidal. It is widely found in nature in such plants as marine seaweeds, where it is bound to chemical compounds. Iodine acts by halogenating tyrosine portions of protein molecules.

*Tincture of iodine*, a commonly used antiseptic for wounds, consists of 2 percent iodine and sodium iodide dissolved in ethyl alcohol. For the disinfection of *clear water*, the CDC recommends 5 drops of tincture of iodine in 1 quart of water, with 30 minutes contact time before consumption. Iodine compounds in different forms are also valuable sanitizers for restaurant eating utensils and equipment.

*Iodophors* are iodine-detergent complexes that release iodine over a long period of time and have the added advantage of not staining tissues or fabrics. The detergent portion of the complex loosens the organisms from the surface and the halogen kills them. Some examples of iodophors are Wescodyne, used in preoperative skin preparations; Ioprep, for presurgical scrubbing; Iosan, for dairy sanitation; and Betadine, for local wounds. Iodophors may also be combined with nondetergent carrier molecules. The best known carrier is povidone, which stabilizes the iodine and releases it slowly. However, compounds like these are not self-sterilizing. In 1989, for example, four cases of peritoneal *Pseudomonas cepacia* infection were related to a contaminated povidone-iodine product.

**Phenol and Phenolic Compounds**

*Phenol* and phenolic compounds (phenolics) have played a key role in disinfection practices since Joseph Lister used them in the 1860s. Phenol remains the standard against which other antiseptics and disinfectants are evaluated in the phenol coefficient test. It is active against Gram-positive bacteria, but its activity is reduced in the presence of organic matter. Biochemists believe that phenol and its derivatives act by coagulating proteins, especially in the cell membrane.

Phenol is expensive, has a pungent odour, and is caustic to the skin, and therefore the role of phenol as an antiseptic has diminished. However, phenol derivatives called *cresols* have greater germicidal activity and lower toxicity than the parent compound. Mixtures of ortho-, meta-, and para-cresol (creosote) are used commercially as wood preservatives for railroad ties, fence posts, and telephone poles.

Combinations of two phenol molecules called *bisphenols* are prominent in modern disinfection and antisepsis. Orthophenylphenol, for example, is used in Lysol, Osyl, Staphene, and Amphyl. Another bisphenol, *hexachlorophene*, was used extensively in the 1950s and 1960s in such products as toothpaste (Ipana), underarm deodorant (Mum), and bath soap (Dial). One product, *pHiso-Hex*, combined hexachlorophene with a pH-balanced detergent cream. Pediatricians recommended it to retard staphylococcal infections of the scalp and umbilical stump, and for general cleansing of the newborn. However, a late 1960s study indicated that excessive amount could be absorbed through the skin

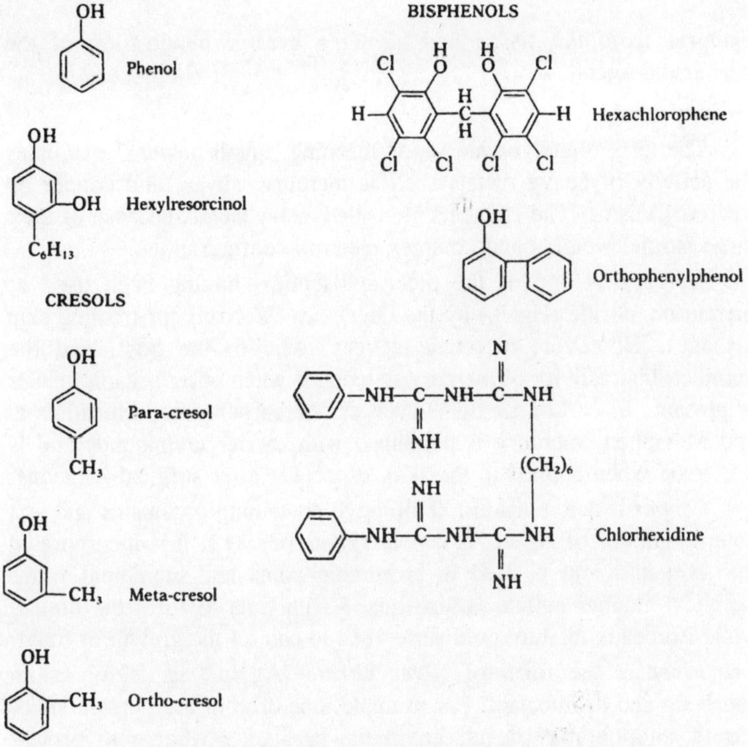

Phenol

BISPHENOLS

Hexachlorophene

Hexylresorcinol

Orthophenylphenol

CRESOLS

Para-cresol

Chlorhexidine

Meta-cresol

Ortho-cresol

*Fig. 8.2. The chemical structures of some important derivatives of phenol used in disinfection and antisepsis.*

and cause neurological damage in newborn rats and monkeys, and hexachlorophene was subsequently removed from over-the-counter products. PHiso-Hex is still available, but only by prescription.

An important bisphenol relative is *chlorhexidine*. This compound was approved in 1976 by the Food and Drug Administration (FDA) for use as a surgical scrub, hand wash, and superficial skin wound cleanser. A 4 percent chlorhexidine solution in isopropyl alcohol is commercially available as *Hibiclens*. The chemical is believed to act on the cell membrane of Gram-positive and Gram-negative bacteria. Chlorhexidine in a concentration of 0.2 percent is also the most extensively tested and most effective antiplaque and antigingivitis agent. However, there is evidence that bacteria may grow within it.

Another phenol derivative, *hexylresorcinol*, is used in a mouthwash and topical antiseptic (ST37) and in throat lozenges (Sucrets). It has the added advantage of reducing surface tension, thereby loosening bacteria from the tissue and allowing greater penetration of the germicidal agent.

**Heavy Metals**

The term oligodynamic action meaning "small power," expresses the activity of heavy metals such as mercury, silver, and copper on microorganisms. The elements are called *heavy metals* because of their large atomic weights and complex electron configurations.

*Mercury* is one of the older antiseptics, having been used as mercuric chloride ($HgCl_2$) by the Greeks and Romans for treating skin diseases. However, mercury is very toxic to the host, and the antimicrobial activity of mercury is reduced when other organic matter is present. In certain products such as Mercurochrome, Merthiolate, and Metaphen, mercury is combined with carrier compounds and is less toxic when applied to the skin, especially after surgical incisions.

*Copper* is active against chlorophyll-containing organisms and is a potent inhibitor of algae. As *copper sulfate* ($CuSO_4$), it is incorporated into algicides and is used in swimming pools and municipal water supplies. Copper sulfate is also mixed with lime to form the bluish-white Bordeaux mixture used since 1882 to control the growth of fungi.

*Silver* is the form of *silver nitrate* ($AgNo_3$) is useful as an antiseptic and disinfectant. For example, one drop of a 1 percent silver nitrate solution may be placed in the eyes of newborns to protect against infection by *Neisseria gonorrhoeae*. The Gram-negative diplococcus can cause blindness if contracted by a newborn during passage through the birth canal. In 1884, Karl S.F. Crede first used silver nitrate to prevent gonococcal eye disease; many states now require that Crede's method be followed (although erythromycin or tetracycline ointments are often substituted since silver nitrate can cause irritation). Silver compounds are also used to treat suturing threads.

Heavy metals are very reactive with proteins, particularly at the protein's sulfhydryl group (—SH), and they are believed to bind protein molecules together by forming bridges between the groups. Because many of the proteins involved are enzymes, the cellular metabolism is disrupted, and the microorganism dies. However, the heavy metals are not sporicidal.

**Alcohols**

*Alcohols* are effective skin antiseptics and are valuable disinfectants for medical instruments. For practical use, the preferred alcohol is ethyl alcohol. *Ethyl alcohol* is active against vegetative bacterial cells, including the tubercle bacillus, but it has no effect on spores. It denatures proteins and dissolves lipids, an action that may lead to cell-membrane disintegration. Ethyl alcohol also is a strong dehydrating agent. Because ethyl alcohol reacts readily with any organic matter, medical instruments and thermometers must be thoroughly cleaned before exposure. Usually, a 50 to 80 percent alcohol solution is recommended since water prevents rapid evaporation and assists penetration into the tissues. A 10-minute immersion in 70 percent ethyl alcohol is generally sufficient to disinfect a thermometer or delicate instrument.

Alcohol is used to preserve cosmetics, and to treat skin before a venipuncture or injection. It mechanically removes bacteria from the skin and dissolves lipids to clean the skin. *Isopropyl alcohol*, or rubbing alcohol, has high bactericidal activity in concentrations as high as 99 percent. Methyl alcohol is toxic to the tissues and is used infrequently.

## Other Chemical Agents

The chemical agents we have discussed in previous sections are best recognized as disinfectants and antiseptics. Augmenting these are some chemical that can be used for sterilization purposes, especially for modern high-technology equipment (as well as the rather mundane Petri dish). Three such agents are considered next.

### Formaldehyde

*Formaldehyde* is a gas at high temperatures and a solid at room temperatures. When 37 grams of the solid are suspended in 100 ml of water, a solution called *formalin* results. For over a century formalin has been used in embalming fluid for anatomical specimens and by morticians as well as for disinfecting purposes. In microbiology, formalin is utilized for inactivating viruses in certain vaccines and producing toxoids from toxins.

In the gaseous form, formaldehyde is expelled into a closed chamber where it is a sterilant for surgical equipment, hospital gowns, and medical instruments. However, penetration is poor, and the surface must be exposed to the gas for up to 12 hours for effective sterilization. Instruments can be sterilized by placing them in a 20 percent solution of formaldehyde in 70 percent alcohol for 18 hours. Formaldehyde, however, leaves a residue, and instruments must be rinsed before use. Many allergic individuals develop a contact dermatitis to this compound.

Formaldehyde is an *alkylating agent*. It reacts with amino and hydroxyl groups of nucleic acids and proteins, and with carboxyl and sulfhydryl groups in proteins by inserting between them a small carbon fragment (an alkyl group) and forming bridges. This insertion changes the structures of the molecules and interferes with an organism's chemistry, thereby leading to death. We shall discuss two other alkylating agents, ethylene oxide and glutaraldehyde, in the next paragraphs.

**Ethylene Oxide**

The development of plastics for use in microbiology required a suitable method for sterilizing these heat-sensitive materials. In the 1950s, research scientists discovered the antimicrobial abilities of *ethylene oxide* (EtO) and essentially made the plastic Petri dish and plastic syringe possible.

Ethylene oxide is a small molecule with excellent penetration capacity and sporicidal ability. However, it is both toxic and highly explosive. Its explosivensss is reduced by mixture with Freon gas in Cryoxide or carbon dioxide gas in Carboxide, but its toxicity remains a problem for those who work with it. The gas is released into a tightly sealed chamber where it circulates for up to 4 hours with carefully controlled humidity. The chamber must then be flushed with inert gas for 8 to 12 hours to ensure that all traces of EtO are removed, otherwise the chemical will cause "cold burns" on contact with the skin.

Ethylene oxide has been employed to sterilize paper, leather, wood, metal, and rubber products, as well as plastics. In the hospital it is used to sterilize catheters, artificial heart valves, heart-lung machine components, and optical equipment. The National Aeronautics and Space Administration (NASA) uses the gas for sterilization of interplanetary space capsules. Ethylene oxide chambers have become chemical counterparts of autoclaves for sterilization procedures. Often they are called gas autoclaves.

A closely related gas, *beta-propiolactone* (BPL), is less explosive than ethylene oxide, but its penetrating capacity is more limited. As a liquid it is used to sterilize vaccines, sera, and surgical ligatures. However, scientists have shown it to be carcinogenic, and BPL is therefore used only under restricted circumstances.

**Glutaraldehyde**

*Glutaraldehyde* has become one of the most effective chemical liquids for sterilization purposes. This small molecule destroys vegetative cells within 10 to 30 minutes and spores in 10 hours. Glutaraldehyde is an alkylating agents, usually employed as a 2 percent solution. To use it for sterilization, materials have to be precleaned, immersed for 10 hours, rinsed thoroughly with sterile water, dried in a special cabinet with sterile air, and stored in a sterile container to ensure that the material remains sterile. If any of these parameters is altered, the materials may be disinfected but may not be considered "sterile." Because the activity of glutaraldehyde is not greatly reduced by organic matter, the chemical is recommended for use on surgical instruments where residual blood may be present. In addition, glutaraldehyde does not damage delicate objects, and therefore it can be used to sterilize optical equipment such as the optic fiber endoscopes used for arthroscopic surgery. It gives off irritating fumes, however, and instruments must be rinsed thoroughly in sterile water. At pH 7.5, glutaraldehyde kills *S. aureus* in 5 minutes and *M. tuberculosis* in 10 minutes.

**Hydrogen Peroxidase**

*Hydrogen peroxide* ($H_2O_2$) is used as a rinse in wounds, scrapes, and abrasions. The area foams and effervesces as *catalase* in the tissue breaks down hydrogen peroxide to oxygen and water. The furious bubbling removes microorganisms mechanically. Hydrogen peroxide decomposition also results in a reactive form of oxygen—the superoxide radical—highly toxic to microorganisms. Anaerobic bacteria are sensitive to hydrogen peroxide because the sudden release of oxygen gas inhibits their growth. New forms of $H_2O_2$ such as Super D Hydrogen Peroxide are more stable than traditional forms and therefore do not decompose spontaneously. Such inanimate materials as soft contact lenses, utensils, heat-sensitive plastics, and food processing equipment can be disinfected within 30 minutes. Sterilization can also be achieved after 6 hours of exposure to a 6 percent solution. Microbiologists are currently researching the use of hydrogen peroxide for destroying bacteria in milk and milk products.

**Soaps and Detergents**

A *soap* is a chemical compound of fatty acids combined with potassium or sodium hydroxide. The pH of the compound is usually about 8.0, and some microbial destruction is therefore due to the alkaline conditions established on the skin. However, the major activity of soap is as a degerming agent for the mechanical removal of microorganisms from the skin surface.

Soaps are *wetting agents*, that is they emulsify and solubilize particles clinging to surface. The surface tension is also reduced by the soap. In addition, soaps remove skin oils, further reducing the surface tension and increasing the cleaning action.

*Detergents* are synthetic chemicals acting as strong wetting agents and surface tension reducers. Since they are actively attracted to the phosphate groups of cellular membranes, they also alter the membranes and encourage leakage from the cytoplasm.

Certain *anionic detergents* yield negatively charged ions in solution. The detergents are somewhat active against Gram-positive bacteria, but the negative charges of bacteria usually repel them and limit their use to common laundry products. Anionic detergents such as Triton W-30 and Duponol find value as iodophors.

Other detergents are *cationic.* These derivatives of ammonium chloride contain four organic radicals in place of the four hydrogens, and at least one radical is a long-chain alkyl group. The positively charged ammonium group is counterbalanced by a negatively charged chloride ion. Such compounds are often called *quaternary ammonium compounds* or simply, quats.

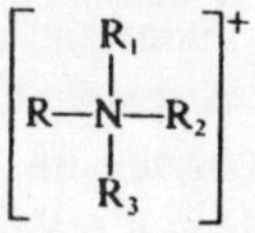

$[R{-}N(R_1)(R_3){-}R_2]^+$ Basic structure of cationic detergents

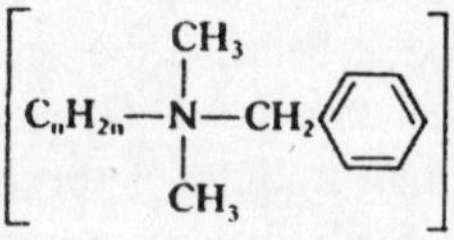

$[C_nH_{2n}{-}N(CH_3)_2{-}CH_2C_6H_5]Cl$ Benzalkonium chloride (Zephiran)

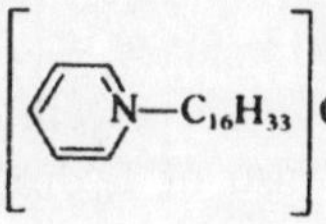

$[C_5H_5N{-}C_{16}H_{33}]Cl$ Cetylpyridinium chloride (Ceepryn chloride)

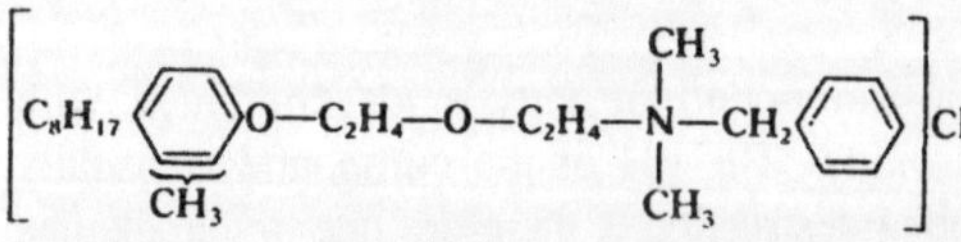

$[C_8H_{17}C_6H_3(CH_3){-}O{-}C_2H_4{-}O{-}C_2H_4{-}N(CH_3)_2{-}CH_2C_6H_5]Cl$ Methylbenzethonium chloride (Diaparene chloride)

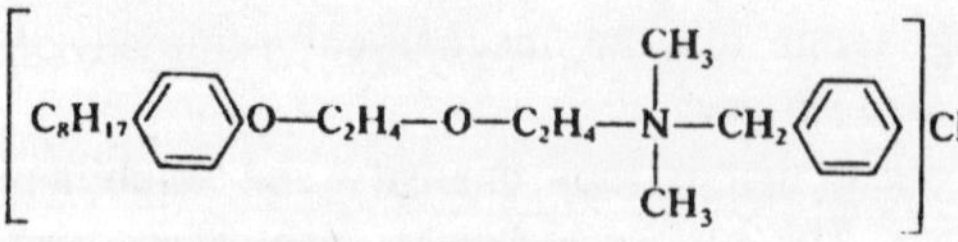

$[C_8H_{17}C_6H_4{-}O{-}C_2H_4{-}O{-}C_2H_4{-}N(CH_3)_2{-}CH_2C_6H_5]Cl$ Benzethonium chloride (Phemerol chloride)

*Fig. 8.3. The chemical structures of some important cationic detergents used in disinfection and antisepsis.*

The compounds in cationic detergents have rather long, complex names, such as benzalkonium chloride in Zephiran and cetylpyridinium chloride in Ceepryn. Other detergents are used in Phemerol and Diaparene. Cationic detergents are bacteriostatic on a broad range of bacteria, especially Gram-positive bacteria, and are relatively stable, with little odour. They are used as sanitizing agents for industrial equipment and food utensils, as skin antiseptics, in mouth-washes and storage solutions for contact lenses, and for disinfecting hospital walls and floors. Mixing with soap, however, reduces their activity, and certain Gram-negative bacteria such as *Pseudomonas aeruginosa* can grow in them.

**Dyes**

Dyes are useful in microbiology as staining reagents and in laboratory media, where they help select out certain organisms from a mixture. A group of dyes called *triphenylmethane dyes* is also useful as antiseptics against species of *Bacillus* and *Staphylococcus*. The group includes malachite green and crystal violet. Crystal violet has also been used traditionally as *gentian violet* for trench mouth and for *Candida albicans* infections such as thrush. Interference with cell wall construction appears to be the mode of activity. The dye is bactericidal at very weak dilution of less than 1:10,000.

A second group of dyes, the *acridine dyes*, includes acriflavine and proflavine. Both dyes are used as antiseptics for staphylococcal infections in wounds. They apparently act by combining directly with DNA, thereby halting RNA synthesis.

**Acids**

Certain acids are useful as antiseptics or disinfectants. The popular ones include benzoic, salicylic, and undecylenic acids for tinea infections of the skin. These infections are caused by various species of fungi.

*Organic acids* are particularly valuable as food preservatives. Lactic and acetic acids, for example, are important preservatives in sour foods such as cheeses, sauerkraut, and pickled products. Propionic acid is added to bakery products to keep microbial populations low.

In general, acid enhances the effects of disinfectants and antiseptics and makes them more soluble. Also, heat is a more potent sterilizing agent if acid conditions are present, because hydrolysis is increased. Acid is therefore a valuable adjunct to disinfection, albeit in an indirect way.

# 9

# ANTIBIOTICS

For centuries, physicians believed that heroic measures were necessary to save proteins from the ravages of infectious disease. They prescribed frightening courses of purges and bloodlettings, enormous doses of strange chemical concoctions, ice water baths, starvation, and other drastic remedies. These treatments probably complicated an already bad situation by reducing the natural body defenses to the point of exhaustion.

But a revolution in medicine took place about 1825 when a group of doctors in Boston and London experimented to see what would happen if such treatments were withheld from diseased patients. Surprisingly, they found the survival rate essentially the same and, in some cases, better. Over the next few decades, the lessons from their experiments spread, and as the worst features of heroic therapy disappeared, doctors adopted a conservative, nonmeddling approach to disease. It became the doctor's job to diagnose the illness, explain it to the family, predict what would happen in the next several days, and then stand by and care for the patient within the limits of what was known.

When the germ theory of disease emerged in the late 1800s, the information about microorganisms added considerably to the understanding of disease and increased the storehouse of knowledge available to the doctor. However, it did not change the fact that little, if anything, could be done for the infected patient. Tuberculosis continued to kill one of every seven people dead from all causes; and streptococcal disease was a fatal experience, as were pneumococcal pneumonia and meningococcal meningitis.

Then, in the 1940s, the chemotherapeutic agents and antibiotics burst on the scene, and another revolution in medicine began. Doctors were astonished to learn that they could kill bacteria in the body without doing substantial harm to the body itself. Medicine experienced a period of powerful, decisive therapy for infectious disease, and doctors found they could successfully alter the course of disease. The chemotherapeutic agents and antibiotics effected a radical change in medicine and charted a new course that has followed through to the present day.

In this chapter we shall discuss the antimicrobial drugs that have become mainstays of our health-care delivery system. We shall explore their discovery and examine their uses, while nothing the important side effects attributed to many of them. When Pasteur performed his experiments a hundred years ago, he implied that microorganisms could be destroyed and that some day, a way would be found to successfully treat many diseases. Only since the 1940s has Pasteur's prophecy become reality.

## Chemotherapeutic Agents

*Chemotherapeutic agents* are chemical substances used within the body for therapeutic purposes. The term generally implies a chemical that has been synthesized by chemists or produced by a modification of a preexisting chemical. By contrast, an *antibiotic* is a product of the metabolism of a microorganism. We shall maintain the disinfection in these pages, although many antibiotics are currently produced by synthetic or semisynthetic means and are more correctly "chemotherapeutic agents." Our discussion of chemotherapeutic agents will begin with a brief review of their development.

### A Short History of Chemotherapy

In the drive to control and cure disease, the efforts of early microbiologists were primarily directed toward enhancing the body's natural defenses. Sera containing antibodies lessened the impact of diphtheria, typhoid fever, and tetanus; and effective vaccines for smallpox and rabies (and later, tuberculosis, diphtheria, and tetanus) reduced the incidence of these diseases.

Among the leaders in the effort to control disease was an imaginative investigator named *Paul Ehrlich*. Ehrlich envisioned antibody molecules as "magic bullets" that seek out and destroy disease organisms in the tissues without harming the tissues. His experiments in stain technology indicated that certain dyes also had antimicrobial qualities, and by the early 1900s, his attention had turned to magic bullets of a purely chemical nature.

One of Ehrlich's collaborators was the Japanese investigator *Sahachiro Hata*. Hata wished to perform research on the chemical control of the syphilis spirochete *Treponema pallidum*, and Ehrlich was happy to oblige. Previously, Ehrlich and his staff had synthesized hundreds of arsenic-phenol compounds, and Hata set to work testing them for antimicrobial qualities. After months of painstaking study, Hata's attention focused on *arsphenamine*, compound #606 in the series. Hata and Ehrlich successfully tested arsphenamine against syphilis in animals and human subjects, and in 1910, they made a derivative of the drug available to doctors for use against the disease. Arsphenamine, the first modern chemotherapeutic agent, was given the common name *Salvarsan* because it offered salvation from syphilis and contained arsenic.

Salvarsan met with mixed success during the ensuring years. Its value against syphilis was without question, but local reactions at the injection site and indiscriminate use by some physicians brought adverse publicity. Moreover, some church official used the threat of syphilis as a deterrent to immoral behaviour, and they were less than enthusiastic about Salvarsan's therapeutic effect. Ehrlich's death in 1915 together with the general ignorance of organic chemotherapy. Instead, interest strengthened in serum and vaccine therapy for war-related wounds and diseases. Significant advances in chemotherapy would not be made for another 20 years. During this interval, German chemist continued to synthesize and manufacture dyes for fabrics and other industries, and they routinely tested their new products for antimicrobial qualities. Among these products was a red dye, *prontosil*, synthesized in 1932. Prontosil had no apparent effect on bacteria in culture.

But things were different in animals. When the German chemist *Gerhard Domagk* tested prontosil in animals he found a pronounced inhibitory effect on staphylococci, streptococci, and other Gram-positive bacteria. In February 1935, Domagk injected the dye to his daughter Hildegrade, who was gravely ill with septicemia. She had pricked her finger with a needle, and blood infection had followed rapidly. Hildegrade's condition gradually improved, and, to many historians, her recovery set into motion the age of modern chemotherapy. For his discovery, Gerhard Domagk was awarded the 1939 Nobel Prize in Physiology or Medicine (*in absentia*, however, because Hitler forbade him to accept it).

Later in 1935, a group at the Pasteur Institute headed by *Jacques and Therese Trefouel* isolated the active principle in prontosil. They found it to be *sulfanilamide*, a substance first synthesized by Paul Gelmo in 1908. Sulfanilamide quickly became a mainstay for the treatment of wound-related infections sustained during World War II.

**Sulfanilamide and Other Sulfonamides**

Sulfanilamide was the first of a group of chemotherapeutic agents known as *sulfonamides*. In 1940, the British investigators *D.D. Woods* and *E.M. Fildes* proposed a mechanism of action for sulfanilamide and other sulfonamides, and gave insight on how they interfere with the metabolism of bacteria without damaging body tissues. The mechanism came to be known as *competitive inhibition*.

According to the mechanism of competitive inhibition, certain bacteria synthesize an important molecule called *folic acid* for use in nucleic acid production. Humans cannot synthesize folic acid and must consume it in foods or vitamin capsules. However, bacteria possess the necessary enzyme to manufacture folic acid and are incapable of absorbing folic acid from the surrounding environment.

In the production of folic acid, the bacterial enzyme joins together three important components, one of which is *para-aminobenzoic acid* (PABA). This molecule is similar to sulfanilamide in chemical structure. Therefore if the environment contains large amounts of sulfanilamide, the enzyme selects the sulfanilamide molecule instead of the PABA molecule for use in folic acid production. Once combined with the enzyme, the molecule binds tightly, and effectively inhibits the enzyme, thus making it unavailable for folic acid synthesis. As the production of folic acid is reduced, nucleic acid synthesis ceases, and the bacteria die.

Modern sulfonamides are typified by *sulfamethoxazole*. Doctors prescribe this drug for urinary tract infections due to Gram-negative rods, and for meningococcal meningitis. Frequently the drug is combined with *trimethoprim*, a drug that inhibits another step in folic acid synthesis. Commercially the drug combination is known as Bactrim. It is frequently used to treat *Pneumoncystis* pneumonia. Another common sulfonamide, *sulfisoxazole*, is marketed as Gantrisin cream for vaginal infections due to Gram-negative bacteria. In some patients, a drug allergy to sulfonamides develops, with a skin rash, gastrointestinal distress, or type II cytotoxic hypersensitivity.

**Other Chemotherapeutic Agents**

The discovery and development of sulfanilamide led to the development of numerous other chemotherapeutic agents, many of which are currently in wide use. Onc example is the antituberculosis drug *isoniazid* (isonicotinic acid hydrazide, INH). Biochemists believe that isoniazid interferes with cell-wall synthesis in *Mycobacterium* species by inhibiting the production of mycolic acid, a component of the wall. Isoniazid is often combined in therapy with such drugs as rifampin and ethambutol. *Ethambutol* is a synthetic, well-absorbed drug that is tuberculocidal. Visual disturbances limit its use to treatment of tuberculosis.

Another chemotherapeutic agent, a quinolone called *nalidixic acid*, blocks DNA synthesis in certain Gram-negative bacteria that cause urinary tract infections. Synthetic derivatives of nalidixic acid called *fluoroquinolones* are also used in urinary tract infections as well as for gonorrhea and chlamydia and for intestinal tract infections due to Gram-negative bacteria. Examples of the fluoroquinolone drugs are ciprofloxacin (Cipro), enoxacin, and norfloxacin.

*Nitrofurantoin* is a drug actively excreted in the urine for urogenital infections. *Metronidazole* (Flagyl) has been used for decades against trichomoniasis, amebiasis, and giardiasis. However, evidence that the drug causes tumours in mice has prompted physicians to prescribe it with caution. The treatment of malaria has long depended upon the consumption of *quinine*. When the tree bark used in its production became unavailable during World War II, researchers quickly set to work to develop two alternatives: chloroquine and primaquine. *Chloroquine* is effective for terminating malaria attacks; *primaquine* destroys the malaria parasites outside red blood cells.

Two other chemotherapeutic drugs, both inhibitory to *Mycobacterium* species, bear brief mention. The first is *para-aminosalicylic acid* (PAS), a drug that closely resembles sulfonamides and is used for tuberculosis. The second agent is diaminodiphenylsulfone, or *dapsone*, used to treat leprosy.

## Antibiotics

The word antibiotics is derived from Greek stems that mean "against life." In 1889, the French researcher *Paul Vuillemin* coined "antibiotic" to describe a substance he isolated some years earlier from *Pseudomonas aeruginosa*. The substance, called pyocyanin, inhibited the growth of other bacteria in test tubes but was too toxic to be useful in disease therapy. Vuillemin's term has survived to the current era. *Antibiotics* are now considered to be chemical products or derivatives of certain organisms that are inhibitory to other organisms.

Scientists are uncertain as to how the ability to produce antibiotics arose in living things, but it is conceivable that random genetic mutations were responsible. Clearly, the ability to produce an antibiotic conferred an extraordinary evolutionary advantage on the possessor in the struggle for survival. In this section we shall discuss the sources of antibiotics, their modes of action, and side effects, and how they are used by physicians to control microorganisms. Our study will begin with Fleming's discovery of penicillin and the events that followed.

**Discovery of Antibiotics**

One of the first to postulate the existence and value of antibiotics was the British investigator, *Alexander Fleming*. Fleming was a student of Almroth Wright, the discoverer of opsonins. During his early years, Fleming experienced the excitement of the Golden Age of Microbiology and spoke up for the therapeutic value of Salvarsan. In a series of experiments in 1921, he described lysozyme, the nonspecific enzyme that breaks down cell walls in Gram-positive bacteria.

The discovery of antibiotics is an elegant expression of Pasteur's dictum: "Chance favours the prepared mind." In 1928, Fleming was performing research on staphylococci at St. Mary's Hospital in London. Before going on vacation, he had spread staphylococci on plates of nutrient agar, and now, on his return, he noted that one plate was contaminated by a green mold. His interest was piqued by the failure of staphylococci to grow near the mold. Fleming isolated the mold, identified it as a species of *Penicillium*, and found that it produced a substance that killed Gram-positive organisms. Though he failed to isolate the substance, he named it *penicillin*.

Fleming was not the first to take note of the antibacterial qualities of *Penicillium* species. Joseph Lister observed a similar phenomenon in 1871, John Tyndall did likewise in 1876, and a French medical student Ernest Duchesne wrote a research paper on the subject in 1896. Whether they were observing the effects of penicillin or some other inhibitor is not known. However, we know that Fleming proposed using penicillin to eliminate Gram-positive bacteria from mixed cultures, and that he unsuccessfully tried the filtered broth on infected wound tissue. At the time, vaccines and sera were viewed as essential to disease therapy, and Fleming's request for financial support went unheeded. Moreover, biochemistry was not sufficiently advanced to make complex separations possible, and funds for research were limited since the Depression had begun. Fleming's discovery was soon forgotten.

In 1935, Gerhard Domagk's dramatic announcement of the chemotherapeutic effects of prontosil fueled speculation that chemicals could be used to fight disease in the body. Then, in 1939, Rene Dubos of the Rockefeller Institute in New York City reported that soil bacteria could produce antibacterial substances. By that time, a group at Oxford University led by pathologist *Howard Florey* and biochemist *Ernst Boris Chain* had reisolated Fleming's penicillin and were conducting trails with highly purified samples. An article in *The Lancet* in 1940 detailed their success. But England was deeply involved in World War II, so a group of American companies developed the techniques for large-scale production of penicillin and made the drug available for commercial use. Fleming, Florey, and Chain shared the 1945 Nobel Prize in Physiology or Medicine for the discovery and development of penicillin.

**Penicillin**

Since the 1940s, penicillin has remained the most widely used antibiotic because of its low cost and thousands of derivatives. *Penicillin G*, or benzylpenicillin, is currently the most popular penicillin antibiotics and is usually intended when the physician prescribes "penicillin." Other types are penicillin F and penicillin V, all with the same basic structure of a *beta-lactam nucleus* and several attached groups.

The penicillins are active against a variety of Gram-positive bacteria, including staphylococci, streptococci, clostridia, and pneumococci. In higher concentrations, they are also inhibitory to the Gram-negative diplococci that cause gonorrhea and meningitis, and are useful against syphilis spirochetes. Penicillin functions during the synthesis of the bacterial cell wall. It blocks cross-linking of carbohydrates in the peptidoglycan layer during wall formation, resulting in such a weak wall that internal pressure causes the cell to swell and burst. Penicillin is therefore bactericidal in rapidly multiplying bacteria (as in an infection). Where bacteria are multiplying slowly or are dormant, the drug may have only a bacteriostatic effect or no effect at all.

Over the years, two major drawbacks to the use of penicillin have surfaced. The first is the *anaphylactic reaction* occurring in allergic individuals. This allergy applies to all compounds related to penicillin. Swelling about the eyes or wrists, flushed or itchy skin, shortness of breath, and a series of hives are signal that sensitivity exists and that penicillin therapy should cease immediately.

The second disadvantage is the evolution of penicillin-resistant bacteria. These organisms produce *penicillinase* (also called beta-lactamase), an enzyme that converts penicillin into harmless penicilloic acid. It is probable that the ability to produce penicillinase has always

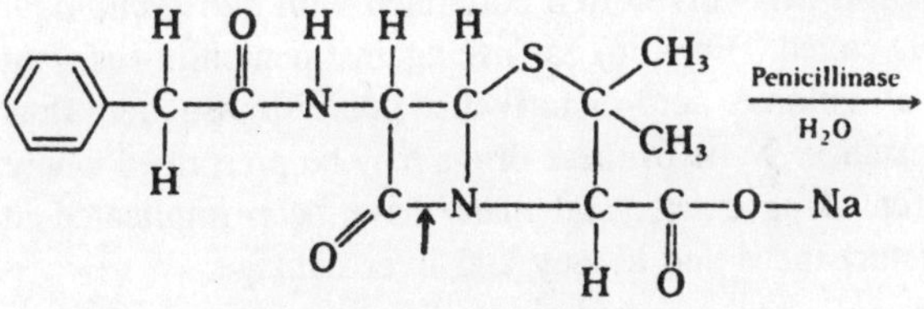

Penicillinase
$H_2O$

Sodium penicillin G

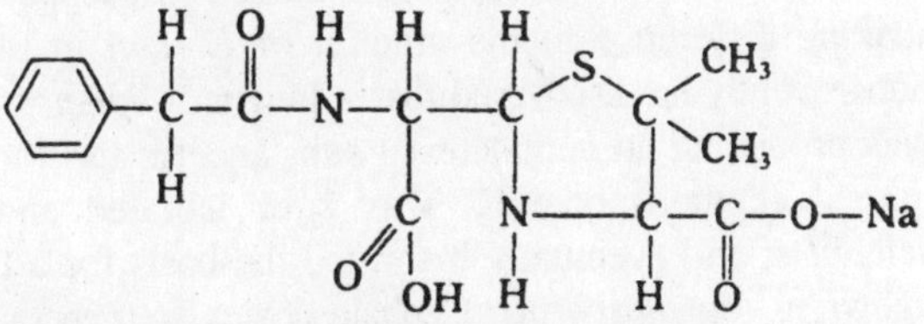

Sodium penicilloic acid

*Fig. 9.1. The action of penicillinase on sodium penicillin G.*

existed in certain bacterial mutants, but that the ability manifests itself when the organisms are confronted with the drug. Thus a natural selection takes place, and the rapid multiplication of penicillinase-producing bacteria yields organisms against which penicillin is useless. Recent years, for example, have witnessed an increase in penicillinase-producing *Neisseria gonorrhoeae* (PPNG), with the result that penicillin is now less useful for gonorrhea treatment.

## Semisynthetic Penicillins

In the late 1950s, the beta-lactam nucleus of the penicillin molecule was identified and synthesized, and it became possible to attach various groups to this nucleus and create new penicillins. In the following years, thousands of penicillins emerged from this semisynthetic process.

*Ampicillin* exemplifies a semisynthetic penicillin. It is less active against Gram-positive cocci than penicillin G, but is valuable against several Gram-negative rods as well as gonococci and meningococci. The drug resists stomach acid and is absorbed from the intestine after oral consumption. *Amoxicillin*, a chemical relative of ampicillin, is also acid-stable and has the added advantage of not binding to food as many antibiotics do. Because ampicillin and amoxicillin are excreted into the urine, they are used to treat urinary tract infections.

Another semisynthetic penicillin, *carbenicillin*, is used primarily for infections of the urinary tract. Other semisynthetic penicillins include methicillin, nafcillin, piperacillin, and oxacillin. Still another is *ticarcillin*, a penicillin derivative often combined with *clavulanic acid* (the combination is called Timentin) for use against penicillin-resistant organisms. The clavulanic acid inactivates penicillinase and thus overcomes the resistance. None of these drugs may be prescribed where allergy to the parent drug exists, and many have been implicated in gastrointestinal disturbances and kidney and liver damage.

**Cephalosporins**

While evaluating sea water samples along the coast of Sardinia in 1945, an Italian microbiologist named Giuseppe Brotzu observed a striking difference in the amount of *E. coli* in two adjoining areas. Subsequently he discovered that a fungus, *Cephalosporium acremonium*, was producing an antibacterial substance in the water. The substance, named cephalosporin C, was later isolated and characterized by scientists, and eventually it formed the basis for a family of antibiotics known as *cephalosporins*. Cephalosporins are generally arranged in three groups or "generations." *First-generation* cephalosporins are variably absorbed from the gut and are useful against Gram-positive cocci and certain Gram-negative rods. They include cephalexin (Keflex) and cephalothin (Keflin). *Second-generation* drugs are active against Gram-positive cocci as well as Gram-negative rods (e.g., *Haemophilus influenzae*) and include cefaclor, cefoxitin, and cefuroxime (Zinacef). The *third-generation* cephalosporins are used primarily against Gram-negative rods (e.g., *Pseudomonas aeruginosa*) and for treating diseases of the central nervous system. Cefotaxime (Claforan), ceftriaxone (Rocephin), and ceftazidime (Fortaz) are in the group.

Cephalosporins resemble penicillins in chemical structure except that the beta-lactam nucleus has a slightly different composition. The antibiotics are used as alternatives to penicillin where resistance is encountered or in cases where penicillin allergy exists. Side effects appear to be minimal but allergic reactions have been reported and thrombophlebitis can occur. The drugs function by interfering with cell wall synthesis in bacteria.

**Aminoglycosides**

The *aminoglycosides* are a group of antibiotics compounds in which amino groups are bonded to carbohydrate molecules (glycosides) bonded to other carbohydrate molecules. All aminoglycosides attach irreversibly to bacterial ribosomes, thereby blocking the reading of the genetic code on messenger RNA molecules. Since oral absorption is negligible, the antibiotics must be administered by injection. Their use has declined in resent years with the introduction of second- and third-generation cephalosporins and with the introduction and development of quinolone drugs such as the fluoroquinolones.

In 1943, the first aminoglycoside was discovered by researchers led by *Selman A. Waksman*, a soil microbiologist at Rutgers University. Waksman's group isolated an antibacterial substance from a moldlike bacterium named *Streptomyces griseus* and named the substance *streptomycin*. At the time, the discovery was sensational because streptomycin was useful against tuberculosis and numerous diseases caused by Gram-negative bacteria. Since then it has been largely replaced by safer drugs, but streptomycin is still prescribed on occasion for such diseases as tuberculosis. The major side effect of therapy is damage to the auditory branch of the nerve extending from the inner ear. Deafness may result.

*Gentamicin*, a widely used aminoglycoside, is administered for serious infections caused by Gram-negative bacteria, especially those in urinary tract infections. The antibiotic is produced by species of *Micromonospora*, a bacterium related to *Streptomyces*. Damage to the kidneys and the hearing mechanism reflects its toxicity.

Neomycin and kanamycin are two older antibiotics of the aminoglycoside group, having been isolated from *Streptomyces* species in 1949 and 1957, respectively. *Neomycin* is sometimes used to control infections of the intestine because it is poorly absorbed, and it is prescribed as an ointment for bacterial conjunctivitis. Commercially, it is available in combination with polymyxin B and bacitracin as Neosporin. *Kanamycin* is used primarily against Gram-negative bacteria in wounded tissue. A derivative of kanamycin called *amikacin* is used for controlling numerous nosocomial diseases and for intestinal diseases. Physicians use *tobramycin* against *Pseudomonas* species, and *paromomycin* against *Entamoeba histolytica*. Both antibiotics are derived from *Streptomyces* species.

### Chloramphenicol

*Chloramphenicol* was the first broad-spectrum antibiotic discovered. Its isolation in 1947 by John Ehrlich, Paul Burkholder, and David Gotlieb was hailed as a milestone in microbiology because the drug was capable of inhibiting a wide variety of Gram-positive and Gram-negative bacteria, as well as several species of rickettsiae and fungi. During the next 30 years, however, physicians tempered their enthusiasm for chloramphenicol as side effects became apparent and new drugs appeared. Nevertheless, chloramphenicol still retains its importance in the treatment of many diseases.

Chloramphenicol is a small molecule that passes into the tissues, where it interferes with protein synthesis in microorganisms. It diffuses into the nervous system and is thus useful for meningitis. It also remains the drug of choice in the treatment of typhoid fever and is an alternative to tetracycline for typhus fever and Rocky Mountain spotted fever. Originally isolated from the waste products of *Streptomyces venezuelae*, chloramphenicol became the first synthetic antibiotic when scientists at the Parke-Davis Company manufactured it from raw materials (its trade name is Chloromycetin). The drug has two major side effects: In the bone marrow it prevents hemoglobin incorporation into the red blood cells and induces a condition called *aplastic anemia*; and it accumulates in the blood of newborns causing a toxic reaction and sudden breakdown of the cardiovascular system known as the *gray syndrome*. For these reactions, chloramphenicol is not used in trivial infections.

## OTHER ANTIBIOTICS

### Tetracyclines

In 1948, scientists at Lederle Laboratories discovered chlortetracycline, the first of the tetracycline antibiotics. This finding completed the initial quarter of "wonder drugs": penicillin, streptomycin, chloramphenicol, and tetracycline.

Modern *tetracyclines* are a group of broad-spectrum antibiotics with a range of activity similar to chloramphenicols. They include naturally occurring chlortetracycline and oxytetracycline isolated from species of *Streptomyces*, and semisynthetic tetracycline, doxycycline, methacycline, and minocycline. All interfere with protein synthesis in microorganisms by binding to ribosomes.

Tetracycline antibiotics may be taken orally, a factor that led to

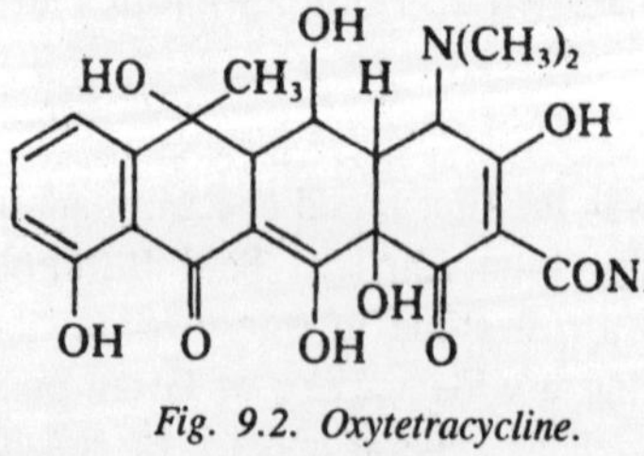

*Fig. 9.2. Oxytetracycline.*

their indiscriminate use in the 1950s and 1960s. The antibiotics were consumed in huge quantities by tens of millions, and in some people, the normal bacterial flora of the intestine was destroyed. With these natural controls eliminated, fungi such as *Candida albicans* flourished. Patients then had to take an antifungal antibiotic such as nystatin, but because this drug was sometimes toxic, the preferred course was to replace the intestinal bacteria by consuming large quantities of bacteria-laden yogurt. Tetracyclines also cause a yellow-gray-brown discolouration of teeth and stunted bones in children. These problems are minimized by restricting use of the antibiotic in pregnant women and children through the teen years.

Despite these side effects, tetracyclines remain the drugs of choice for most rickettsial and chlamydial diseases, including the STD chlamydia. They are used against a wide spectrum of Gram-negative bacteria, and they are valuable for treating primary atypical pneumonia, syphilis, gonorrhea, pneumococcal pneumonia, and certain protozoal diseases as well as acne. Although resistances have occurred, newer tetracyclines such as minocycline (Minocin) and doxycycline (Vibramycin) appear to circumvent these. There is evidence that tetracycline may have been present in the food of ancient people.

**Miscellaneous Antibiotics**

A miscellaneous group of antibiotics merits brief discussion because the drugs in this group are commonly used in modern therapy.

*Erythromycin* is a clinically important antibiotic in the group of substances called the macrolides. *Macrolides* consist of large carbon rings attached to unusual carbohydrate molecules. In the 1970s, researchers discovered that erythromycin was effective for treating primary atypical pneumonia and Legionnaires' disease. The antibiotic is a *Streptomyces* product and a protein synthesis inhibitor. It is recommended for use against Gram-positive bacteria in patients with penicillin allergies and against both *Neisseria* and *Chlamydia* species that can affect the eyes of newborns. It has few side effects.

Another macrolide antibiotic is *clarithromycin*, a semisynthetic drug. Clarithromycin (Biaxin) acts by binding to ribosomes to inhibit protein synthesis in Gram-negative bacteria and the same variety of gram-positive bacteria as erythromycin. Another macrolide called *azithromycin* (Zithromax) has a similar mode of action and spectrum of activity. Both antibiotics are believed dangerous to fetal tissue and should not be taken by pregnant women.

*Vancomycin*, a cell-wall inhibitor, is a waste product of a *Streptomyces* species. It is administered by intravenous injection against diseases due to Gram-positive bacteria, especially severe staphylococcal diseases where penicillin allergy or bacterial resistance is found. Its major side effects are damage to the ear and kidney; the drug is not routinely prescribed for trivial situations.

*Rifampin* is a semisynthetic drug prescribed (in combinations with isoniazid and ethambutol) for tuberculosis and leprosy patients. It is also administered to carriers of *Neisseria* and *Haemophilus* species that cause meningitis and as a prophylactic when exposure has occurred. It acts by interfering with RNA synthesis in bacteria. Rifampin therapy may cause the urine, feces, tears, and other body secretions to assume an orange-red colour and may cause liver damage. The drug is administered orally and is well absorbed.

*Clindamycin* and its parent drug, *lincomycin*, are alternatives where penicillin resistance is encountered. Both are active against Gram-positive bacteria, including several anaerobic species (e.g., *Bacteroides* species). Use of the antibiotics is limited to serious infections, however, because the drugs eliminate competing organisms from the intestine and permit *Clostridium difficile* to overgrow the area. The clostridial toxins may then induce a condition called *pseudomembranous colitis*, in which membranous lesions cover the intestinal wall.

Both bacitracin and polymyxin B are polypeptide antibiotics produced by *Bacillus* species. These antibiotics are generally restricted to use on the skin because internally they may cause kidney damage and they are poorly absorbed from the intestine. *Bacitracin* is available in pharmaceutical skin ointments and is effective against Gram-positive bacteria such as staphylococci. *Polymyxin B* is valuable against *Pseudomonas aeruginosa* and other Gram-negative bacilli, particularly those that cause superficial infections in wounds, abrasions, and burns. The two antibiotics are often combined with neomycin in Neosporin. Bacitracin inhibits cell wall synthesis, while polymyxin B injures bacterial membranes.

*Spectinomycin* is a *Streptomyces* product that shares chemical properties with the aminoglycosides. The antibiotic came into prominence in the 1970s for use against gonorrhea caused by penicillin-resistant gonococci. It is given by intramuscular injection and appears to interfere with protein synthesis.

*Monobactams* are a group of antibiotics first synthesized by researchers at Squibb Laboratories in the early 1980s. The core of monobactam antibiotics is a beta-lactam nucleus isolated from *Chromobacter violaceum*, a purple-pigmented bacterium. One antibiotic, *moxalactam*, is active against a broad variety of Gram-negative bacteria, especially those involved in nosocomial diseases and bacterial meningitis. Resistance to penicillinase appears to be high but interference with platelet function and severe bleeding have limited its use. Another set of beta-lactam drugs are the *carbapenems*. The representative of this group called *imipenem* (Primaxin) has good activity against a variety of Gram-positive bacteria and Gram-negative rods as well as anaerobes (e.g., *Bacteroides fragilis*). It is often prescribed where resistances occur, and it appears to have minimal side effects.

### Antifungal Antibiotics

Fungal diseases pose a special problem for the medical mycologist because there are few drugs available for treatment. For infections of the intestine, vagina, or oral cavity due to *Candida albicans*, the recommended drug is usually *nystatin*. This product of *Streptomyces* is commercially available as Mycostatin or Achrostatin and is sold in ointment, cream, or suppository form. It acts by changing the permeability of the cell membrane by combining with fungal sterols. Often it is combined with antibacterial antibiotics to retard *Candida* overgrowth of the intestine during the treatment of bacterial diseases. Another antifungal product of *Streptomyces*, candicidin, has similar uses but is not as widely used.

*Griseofulvin* is an antibiotic used for fungal infections of the skin, hair and nails such as ringworm and athlete's foot. Griseofulvin interferes with mitosis and causes the tips of molds to curl. It is a product of a *Penicillium* species and is taken orally. For serious systemic fungal infections, the drug of choice is *amphotericin B*. This antibiotic degrades the cell membranes of fungal cells, and is effective for treating serious diseases. However, it causes a wide variety of side effects and therefore is used only in progressive and potentially fatal cases.

Other antifungal antibiotics are synthetic compounds. One example, *flucytosine*, is converted in fungal cells to an inhibitor that interrupts nucleic acid synthesis. The drug is used primarily with amphotericin B in systemic diseases. Another example, the *imidazoles*, include clotrimazole, miconazole, itraconazole, and ketoconazole. These compounds interfere with sterol synthesis in fungal cell membranes. Clotrimazole (Gyne-Lotrimin) is used topically for *Candida* skin infections, while the other drugs are used topically as well as internally for systemic diseases. Side effects are uncommon. Miconazole is commercially available in Micatin for athlete's foot and Monistat 7 for yeast disease. Itraconazole is sold as Sporonox for athlete's foot.

## Antibiotic Assays and Resistance

### Antibiotic Sensitivity Assays

Antibiotics sensitivity assays are used to study the inhibition of a test organism by one or more antibiotics or chemotherapeutic agents.

Two general methods are in common use: the tube dilution method, and the agar diffusion method.

The *tube dilution method* determines the smallest amount of antibiotic necessary to inhibit a test organism. This amount is known as the *minimum inhibitory concentration* (MIC). To determine it, the microbiologist prepares as set of tubes with different concentrations of a particular antibiotic. The tubes are then inoculated with the test organism, incubated, and examined for the growth of bacteria. The extent of growth diminishes as the concentration of antibiotic increases, and eventually an antibiotic concentration is observed at which growth fails to occur. This is the MIC.

The second method, the *agar diffusion method*, operates on the principle that antibiotics will diffuse from a paper disk or small cylinder into an agar medium containing test organisms. Inhibition is observed as a failure of the organism to grow in the region of the antibiotic. A common application of the agar diffusion method is the *Kirby-Bauer test* named after *W.M. Kirby* and *A.W. Bauer*, who developed it in the 1960s. This procedure determines the sensitivity of an organism to a series of antibiotics and is performed according to standards established by the Food and Drug Administration (FDA).

An agar medium such as Mueller-Hinton agar is poured into the plate and inoculated with the organism. Paper disks containing known concentrations of antibiotics are applied to the surface, and the plate is incubated. The appearance of a zone of inhibition surrounding the disk indicates sensitivity. By comparing the diameter of the zones to a standard table, one can determine whether the test organism is susceptible or resistant to the antibiotic. If the organism is susceptible, it will be killed in the patient's bloodstream if the experimental concentration of antibiotic is reached. Resistance indicates that the antibiotic will not be effective at that concentration in the circulation.

**Antibiotic Resistance and Abuse**

During the past 25 years, an alarming number of bacterial species have evolved with resistance to chemotherapeutic agents and antibiotics. Public health microbiologists note that resistant organisms are increasingly responsible for human diseases of the intestinal tract, lungs, skin, and urinary tract. Those in intensive care units and burn wards are particularly vulnerable, as are children, the elderly, and the infirm. Common diseases like bacterial pneumonia, streptococcal sore throat, and gonorrhea that a few years ago succumbed to a single dose of antibiotics are now among the most difficult to treat.

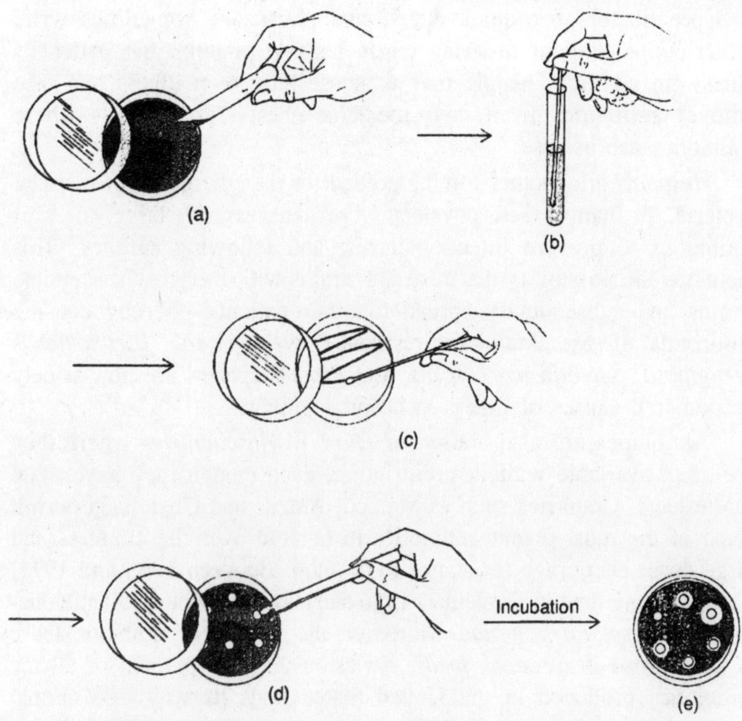


*Fig. 9.3. The Kirby-Bauer test for determining antibiotic susceptibility of an organism by the agar diffusion method.*

Microorganisms may *acquire resistance* to antibiotics in a number of ways. In some cases resistance arises from the microorganism's ability to destroy the antibiotic. The production of penicillinase by penicillin-resistant gonococci is an example. Other resistances are traced to changes in the permeability of the microbial cell wall and membrane, thus prohibiting passage of the antibiotic to the interior. In addition, resistance to the drug's activity may develop. An example of the latter takes place when sulfa drugs fail to unite with enzymes that synthesize folic acid because the enzyme's structure has changed. Moreover, drug resistance may be due to an altered metabolic pathway in the microorganism, a pathway that bypasses the reaction normally inhibited by the drug. An altered structural target for the drug may also evolve. For example, the structure of a pathogen's ribosome may change and make a drug that unites with the ribosome useless.

Resistance may develop in bacteria during the normal course of events, but antibiotic abuse encourages the emergence of resistant forms.

For example, drug companies promote antibiotics heavily, patients pressure doctors for quick cures, and physicians sometimes write prescriptions without ordering costly tests to pinpoint the patients's illness. In addition, people may diagnose their own illness and take leftover antibiotics from their medicine chests for ailments where antibiotics are useless.

*Hospitals* are another forcing ground for the emergence of resistant bacteria. In many cases, physicians use unnecessarily large doses of antibiotics to prevent infection during and following surgery. This increases the possibility that resistant strains will overgrow susceptible strains and subsequently spread to other patients, thereby causing nosocomial disease. Antibiotics-resistant *Escherichia coli*, *Pseudomonas aeruginosa*, *Serratia marcescens*, and *Proteus* species are now widely encountered causes of illness in hospital settings.

Antibiotics are also abused in *Third World countries* where they are often available without prescription, even though they have toxic side effects. Countries such as Mexico, Brazil, and Guatemala permit some of the most potent antibiotics to be sold over the counter, and large doses encourage resistance to develop. Between 1968 and 1971, 12,000 people died in Guatemala from shigellosis attributed to antibiotic-resistant *Shigella dysenteriae*. Moreover, the problem of antibiotic abuse is widespread in *livestock feeds*. An astonishing 40 percent of all the antibiotics produced in the United States finds its way into animal feeds to check disease and promote growth. By killing off less hardy bacteria, chronic low doses of antibiotics create ideal growth environments for resistant strains. Transferred to humans through meat, these resistant organisms may cause intractable illness. An outbreak of salmonellosis in 1984 is illustrative.

Allied to the problem of antibiotic resistance is the concern for transfer of the resistance. Researchers have amply demonstrated that plasmids and transposons accounts for the movement of antibiotic-resistant genes among bacteria. Thus the resistance in a relatively harmless bacterium may be passed to a pathogenic bacterium where the potential for disease is supplemented by resistance to standardized treatment. Antibiotics have traditionally been known as miracle drugs, but there is a growing body of evidence that they are becoming overworked miracles. Some researchers suggest that antibiotics should be controlled as strictly as narcotics. The antibiotic roulette that is currently taking place should be a matter of discussion to all individuals concerned about infectious disease, be they scientist or student.

# 10

# Biological Challenge

The great achievements of molecular biology and genetics over the last 50 years have produced advances in agriculture and industrial processes and have revolutionized the practice of medicine. The very technologies that fueled these benefits to society, however, pose a potential risk as well—the possibility that these technologies could also be used to create the next generation of biological weapons. Biotechnology represents a "*dual use*" dilemma in which the same technologies can be used legitimately for human betterment and misused for bioterrorism.

Events over the 1990s focused growing attention on this balance of risks and benefits, part of a larger concern about the proliferation of weapons of mass destruction (WMD)—chemical, nuclear, or biological. In early 1992, President Yeltsin acknowledged that, despite being an original signatory and State party to the Biological and Toxin Weapons Convention (BWC), the Soviet Union had maintained a major clandestine biological weapons program into the early 1990s. Yeltsin ordered the program shut down, but concerns about other possible secret programs remained. Policymakers in the United States became increasingly concerned that so-called "*rogue states*" would turn to WMD to counter the overwhelming U.S. conventional military superiority. Secretary of Defense Les Aspin launched the "Defense Counterproliferation Initiative" in December 1993 to develop additional means to address these threats. Official statements continue to cite at least a dozen countries believed to have or to be pursuing a biological weapons capability. U.S. and British concerns about Iraq's reported biological and other WMD programs were a primary reason for launching preemptive military action to find and destroy these weapons capabilities. The terrorist attacks of September 11, 2001 and the subsequent anthrax letters accelerated already existing concerns that terrorists would seek WMD capabilities as well. President Bush, in a speech at West Point in 2002, said: "The gravest danger to freedom lies at the perilous crossroads of radicalism and technology. When the spread of chemical and biological and nuclear weapons, along with ballistic missile technology—when that occurs, even weak states and small groups could attain a catastrophic power to strike great nations." States, groups, and individuals are pursuing a biological weapons capability—and the means for them to do so are widely available.

Biological weapons have long been stigmatized as "indiscriminant agents of unnecessary suffering, [whose] use ... contradict(s) the universal principles of war." As discussed below, since November 1969 the U.S. programs linked to biological weapons have been restricted to research and development on defensive measures only. Thus few biologists in the United States today have knowledge of our country's past offensive weapons programs or of the concerns of the national security branches of government. In this respect the life sciences community is in a different situation from that of the physics community, which in large part has been continuously involved in government- sponsored weapons research programs since at least World War II. The scientific community and the government jointly face a double challenge: (1) to establish a working relationship with the national security branches of government, and (2) to help craft a system that will minimize the risk of wrongful use of biological agents or technology without damaging the scientific infrastructure that has made biological research so vital to the health of the nation.

## Life Sciences Today

The biological sciences have experienced enormous growth over the last century, fueled by a stream of discoveries—such as the principles of genctics, the structure of DNA, and the discovery of gene splicing technologies. These have opened new fields of inquiry and provided the basis for myriad applications in industry, agriculture, and medicine. Among the technological breakthroughs in the life sciences, genetic engineering plays a particularly significant role.

Genetic engineering is a technique that permits the artificial modification and transfer of the genetic material from one organism to another and from one species to another. This technology is used throughout the world to alter the protein produced by a gene and to design organisms with desirable traits for applications ranging from basic research and development activities to pharmaceutical and industrial uses. During the last 30 years, these recombinant techniques have spawned a vibrant biotechnology industry focused largely on the development of new pharmaceuticals to fight disease. By 2000 the annual investment in the biotechnology industry peaked at nearly $29 billion, while employment in the biotechnology industry reached 191,000 by 2001.

In response to the opportunities presented by these developments the resources devoted to the life sciences have increased dramatically, making further discoveries possible. The government has funded biological research generously through the National Institutes of Health and National Science Foundation budgets, with few strings attached; private foundations and the pharmaceutical industry have also made major contributions. The number of PhDs awarded each year in the biological and agricultural sciences has increased steadily; 6,526 were awarded in 2001.

This ever-expanding research activity has resulted in numerous new biopharmaceutical products that are transforming medicine. Examples include human recombinant insulin for the treatment of diabetes, a vaccine against hepatitis B, and medicines for diabetes, cancer therapy, arthritis, multiple sclerosis, cystic fibrosis, heart attacks, hemophilia, and sepsis. As knowledge of the human genome increases, it may even become possible to tailor pharmaceutical products not only to specific diseases but also to specific individuals. Throughout this process, the time between new discoveries and their applications has grown ever shorter. One example is the very short time it took the scientific community to identify the coronavirus as the causal agent of the newly emerging human disease, severe acute respiratory syndrome (SARS).

Biotechnology research is now a truly global enterprise. While industrialized countries such as the United States, the United Kingdom, Germany, Israel, and Japan may be the first to develop advanced research and technologies, other countries have a skill base that will enable broad domestic utilization of biological technologies. For example:

China has an aggressive program in plant biotechnology, and as of 2002 plans to increase funding by 400 percent by 2005. This energetic investment also exists in the Chinese private sector, and the national scientific establishment is attempting to lure foreign-trained scientists to return with lucrative financial packages. India is in the process of tripling funding to its national biotech center, and is promoting the development and use of genetically modified crops throughout Asia. Singapore has for many years made a practice of recruiting foreign scientists. Taiwan is investing large amounts in biotechnology and is seeking citizens to return home to build up biotechnology in academia and industry. A Brazilian coalition recently demonstrated sophisticated domestic use of biological technologies by successfully sequencing the plant pathogen *X. fastidiosa* in 2000.

In addition to the dispersed research enterprise, publications and personnel are also widely spread. Well over 10,000 journals in the life sciences are published worldwide. Biological Abstracts, an international database on biology, clinical and experimental medicine, biochemistry, and biotechnology, provides coverage of over 6,000 active international journals and 14,000 archival titles from over 100 countries; Medline, the online service of the National Institutes of Medicine, provides abstract information for more than 4,600 biomedical journals published in the United States and 70 other countries; and PubMed currently provides full-text web access to 4,058 journals. According to Medline, the total number of scientific articles published in the peerreviewed biomedical literature has increased from 449,109 in 1998 to 491,620 in 2001. Given the global nature of the biotechnology research and development enterprise, it is unrealistic to think that biological technologies and the knowledge base upon which they rest can somehow be isolated within the borders of a few countries.

The rapid advance of scientific knowledge and applications owes much to a research culture in which knowledge and biological materials are shared among scientists and people move freely between universities, government agencies, and private industry. Large numbers of foreign graduate students and postdoctoral associates have been an essential ingredient in the success of the biological research enterprise. The scientific workforce is increasingly international; at the National Institutes of Health, for example, approximately 50 percent of the technical staff are non-U.S. citizens. Research results have been widely disseminated, so that even high school students now routinely perform experiments involving recombinant DNA techniques. In short, a dynamic national and international research enterprise has evolved, with an extraordinary record of achievement at multiple centers of excellence. These are values that should be preserved in any sensible policy for minimizing the risks associated with the misapplication of the fruits of the biotechnology enterprise.

## Dual Use Dilemma

The regulation of dual use biotechnology research is a highly contentious technical, political, and societal issue. In the language of arms control and disarmament, dual use refers to technologies intended for civilian application that can also be used for military purposes. Technology involves more than just products; it also encompasses a means to produce and use products in such a way as to solve a problem. Thus, technology comprises "the ability to recognize technical problems, the ability to develop new concepts and tangible solutions to technical problems, the concepts and tangibles developed to solve technical problems, and the ability to exploit the concepts and tangibles in an effective way."

The "*general purpose clause*" of the BWC prohibits the development, production, and stockpiling of biological weapons, but permits States that are parties to the treaty to conduct research activities for peaceful purposes or in order to defend or protect against BW agents. Useful distinctions between permitted and prohibited activities at the level of basic research are difficult to make because biotechnology presents a classic example of the dual use dilemma. In the life sciences, for example, the same techniques used to gain insight and understanding regarding the fundamental life processes for the benefit of human health and welfare may also be used to create a new generation of BW agents by hostile governments and individuals. For the scientists and technicians involved in cutting-edge research and development in biology, biotechnology, medicine, and agriculture, this duality creates both uncertainties and ethical dilemmas. The duality between the purposes permitted and prohibited under the BWC is at the heart of this Committee's activities.

Current research programs in universities, government laboratories, and pharmaceutical companies include experiments directed toward such goals as discovering vaccines for major diseases such as influenza, AIDS, and cancer; new antibiotics for both bacterial and fungal diseases; new sources of genes to protect crops against pests and diseases; and treatments for diabetes, stroke, and Alzheimer's disease. These research activities also include an intense effort to discover vaccines, antibiotics, and detection systems that would provide the defense against each of the select agents. But many of the same methods for developing attenuated live vaccines against viral diseases can have offensive applications as well. The key issue is whether the risks associated with misuse can be reduced while still enabling critical research to go forward.

## Brief History of Modern Biological Warfare

Of thousands of species of potentially pathogenic microorganisms, very few have been developed and deployed as biological weapons. As a society, we tend to think that biological and chemical warfare are recent threats to individuals and populations, but in reality, the offensive use of chemical and biological agents has its origins in antiquity (see Annex to this chapter). It has only been within the last century, however, that infectious disease agents have been seriously considered, on a continuing basis, as tools of war. Based on scientific discoveries during the late nineteenth and early twentieth centuries, biologists were able for the first time to identify, isolate, and culture disease-causing microbes under controlled conditions and use them to intentionally induce disease in a "*naive*" host. "The foundations of microbiology pioneered by Louis Pasteur and Robert Koch offered new prospects for those interested in biological weapons because it allowed agents to be chosen and designed on a rational basis."

Germany was accused of using disease-causing germs during World War I by infecting horses and mules with glanders—a highly infectious animal disease—and cattle with anthrax. German spies were caught in 1917 allegedly trying to spread anthrax bacteria among reindeer herds in the far northern portion of Norway, near the border with Russia. These charges were confirmed when anthrax-laced sugar cubes—obtained from a Swedish-German-Finnish aristocrat arrested as a German agent in 1917—were found to be still viable after being stored in the archives of a Norwegian museum for the last 80 years.

Over the past 60 years pathogens have been identified and perfected as strategic and tactical weapons. Every major combatant during World War II—including the United States, Great Britain, Canada, France, the former Soviet Union, Germany, and Japan—had some type of biological weapons program. During the Sino-Japanese War (1937-1945), Japan repeatedly attacked China with the plague-causing bacteria *Yersinia pestis*, targeting some eleven cities. At least 700 Chinese reportedly died from plague alone, although the number of Chinese civilians killed between 1933 and 1945 by Japanese germ warfare may be much higher.

Japan's secret biological warfare program, Unit 731, officially referred to as the Army Anti-Epidemic Prevention and Water Supply Unit, was located in a remote, highsecurity area in Japanese-occupied Manchuria, first in Harbin and then in Ping Fan. The Japanese perfected culture and dispersal techniques for a large number of biological agents.

After the war the Japanese commander of Unit 731, General Shiro Ishii, traded research data, at the suggestion of his debriefers, with the American occupation government in Japan, in exchange for a grant of immunity from war crimes prosecution. Information obtained from General Ishii later found its way to Camp Detrick, and is still held in the National Archives in the United States.

The United States' offensive biological weapons program also had its origins in World War II. Begun in 1942 within the Chemical Warfare Service at Camp Detrick in Frederick, Maryland, the program's primary mission during World War II was biological warfare research on the causative agents of anthrax and botulism. The main element for carrying out this program, the Special Projects Division of the Army Chemical Warfare Service, had at its peak 3,900 personnel, of which 2,800 were Army, nearly 1,000 Navy, and the remaining 100 civilian. The work was carried out at four installations. Camp Detrick was the parent research and pilot plant center. Field testing facilities were established in 1943 and 1944 in Mississippi and Utah, respectively, and a production plant was constructed in Indiana in 1944 All work, which was coordinated with Great Britain and Canada, was conducted under strictest secrecy.

From the end of World War II until the U.S. decision to renounce its biological weapons program in 1969, this program developed and perfected offensive weapons capabilities for the Air Force, Navy, and the Central Intelligence Agency, utilizing a variety of human, animal, and plant pathogens. "Between 1941 (sic) and 1969, the policy of the United States regarding biological warfare was first (to) deter its use against the United States and its forces, and secondly to retaliate if deterrence failed."

The largest biological weapons complex ever created was in the former Soviet Union. Two main groups of facilities were involved in the research and development, production, and testing of biological weapons: a military-controlled system, which started in the 1920s, and Biopreparat, a top-secret program operating under civilian cover from 1972 until at least 1992, despite the fact that the Soviet Union was an original signatory to and repository for the Biological Weapons Convention. As a result, the Soviet program not only caught up with but surpassed the U.S. program to become the most sophisticated biological weapons program in the world. Its size and scope were enormous; by the early 1990s more than 60,000 people were involved in the research, development, and production of biological weapons as well as the stockpiling of hundreds of tons of anthrax spores and tens of tons of other pathogens, including smallpox and plague. In addition, it is now known that other state programs were involved in aspects of this effort including those of the Ministry of Health, Ministry of Agriculture, Ministry of Defense, KGB, and the Soviet Academy of Sciences.

## U.S. Policy and the Creation of the Biological and Toxin Weapons Convention

After intensive debate in the United Nations and domestic interagency review, President Richard Nixon on November 25, 1969 renounced the first use of lethal and incapacitating chemicals and stated that he would seek ratification of the Geneva Protocol by the U.S. Senate. (The Geneva Protocol of 1925 prohibits the use of chemical or biological materials in war, although it does not proscribe their acquisition or possession.) President Nixon also renounced the use of lethal bacteriological (biological) agents and weapons as well as all other methods of biological warfare, and directed the Defense Department to make recommendations for the disposal of existing BW stockpiles. He further stated that the United States would confine its biological agent and toxin research to defensive measures, such as immunization and safety. On February 14, 1970, this policy was extended to biological toxins regardless of their means of production.

The United States decided to abandon its offensive biological weapons program, destroy its existing stockpiles of biological and toxin weapons, and convert the production facilities to other purposes because it was recognized that:

1. Biological weapons could be as great a threat to large populations as nuclear weapons and that no reliable defense is likely;
2. Biological weapons could be much simpler and less expensive than nuclear weapons to develop and produce; proliferation of biological weapons would therefore greatly increase the number of nations to which the populations of the United States and its allies [could] be held hostage
3. Our biological weapons program was pioneering an easily duplicated technology and was likely to inspire others to follow suit.

The United States concluded that its biological weapons program was a substantial threat to its own national security and that one of the best ways to reduce this threat was not only to renounce biological weapons in this country but also to strengthen the international barriers to their proliferation. The United States, the United Kingdom, and the former Soviet Union together were responsible for the effort to sponsor the Biological and Toxin Weapons Convention (BWC) of 1972—the first arms control agreement to ban outright an entire class of weapons. The U.S. Senate ratified the BWC in 1975. To date 162 countries have signed and 148 countries have ratified the BWC.

## New Threat

The revolution in biotechnology was just beginning when the BWC went into force in 1975. With the advent of the biotechnology revolution and the apparent proliferation of countries desiring to have a biological weapons capability, its signatories must reexamine the efficacy of the Convention in governing the use of disease as a method to spread terror.

The acquisition of biotechnology and biological weapons capability is considerably easier than was the case in the 1940s and 1950s. The explosion in biotechnologies and genetic engineering technologies—all of which have legitimate civilian applications—could empower a hostile agent. Gordon Oehler, Director of the Non-Proliferation Center at the Central Intelligence Agency, testified before the Senate Armed Services Committee, March 27, 1996, and stated that there was "a continuing pursuit by many countries to acquire chemical and biological weapons and that [t]he chilling reality is that these materials and technologies are more accessible now than at any other time in history."

The information to conduct genetic engineering research and advanced technologies are easily accessible on the Internet. Moreover, the equipment and expertise to use this information to create novel agents are available globally. The international diffusion of knowledge and capabilities in biotechnology means that the capacity to carry out beneficial as well as harmful research activities is widely accessible, both to nations and to terrorist groups.

In this situation it is futile to imagine that access to dangerous pathogens and destructive biotechnologies can be physically restricted, as is the case for nuclear weapons and fissionable materials. The nature of the biotechnology problem—indeed the nature of the biological research enterprise—is vastly different from that of theoretical and applied nuclear physics in the late 1930s. The contrast between what is a legitimate, perhaps compelling subject for research and what might justifiably be prohibited or tightly controlled cannot be made a priori, stated in categorical terms, nor confirmed by remote observation.

Matthew Meselson, a leading molecular biologist, gave a stark warning of the potential dangers posed by the destructive applications of biotechnology in May 2000:

Every major technology—metallurgy, explosives, internal combustion, aviation, electronics, nuclear energy—has been intensively exploited, not only for peaceful purposes but also for hostile ones. Must this also happen with biotechnology, certain to be a dominant technology of the coming century? During the century just begun, as our ability to modify fundamental life processes continues its rapid advance, we will be able not only to devise additional ways to destroy life but ... also ... to manipulate it—including the processes of cognition, development, reproduction, and inheritance. A world in which these capabilities are widely employed for hostile purposes would be a world in which the very nature of conflict has radically changed. Therein could lie unprecedented opportunities for violence, coercion, repression, or subjugation.

These dangers cannot be eliminated entirely since the fundamental knowledge from which they emerge is available around the world and the potential benefits of biotechnology for health promotion and national defense are too great to contemplate efforts to prohibit or reverse such research. But the potential adverse effects associated with the malicious exploitation of these technological advances cannot be ignored. Because of widespread moral repugnance against the production and use of chemical and biological weapons (CBW), the involvement of scientists and engineers in CBW research, development and production is widely condemned. History demonstrates, however, that without any military application in mind, research in biology may still contribute to the production of biological weapons. As discussed earlier, the discovery and elaboration of the "*germ theory of disease*" in the nineteenth century led not only to better sanitation and hygiene practices but also to the intentional development of disease as a weapon in the twentieth century.

In discussing modifications of microorganisms that might have significance for bioweapons, Nixdorff and Bender identified four classes of microbial manipulations that have been the subject of intense debate within and outside the scientific community:

1. The transfer of antibiotic resistance to microorganisms,
2. Modification of the antigenic properties of microorganisms,
3. Modification of the stability of microorganisms to the environment, and
4. The transfer of pathogenic properties to microorganisms.

Regarding these manipulations, they observe that: All four kinds of manipulations are possible and are being carried out daily in research laboratories. Some of the most intensive research concerns the elucidation of the mechanisms of pathogenesis. This work is essential for combating infectious diseases. It is hoped that the production of more effective vaccines with [fewer] side effects, better diagnostics and new therapeutic drugs will result from this research. At the same time, it is feared that the advances in biotechnology can be misused to develop and produce biological weapons.

The National Institute of Health's (NIH's) recently released research priorities for countering bioterrorism identified several categories of research activities in immunology and genomics that would be considered "*provocative*" if conducted by a hostile or rogue government. These include efforts to "identify pathogen-induced immunoregulatory and immunosuppressive effects" as well as to "analyze gene expression of agents of bioterrorism." John Gannon, former chairman of the National Intelligence Council and a former deputy director for intelligence at the CIA, observed that "the continuing revolution in science and technology will accentuate the dual use problem related to biotech breakthroughs in biomedical engineering, genomic profiling, genetic modification, and drug development.... Responsible scientists will have an extraordinary opportunity to improve the quality of human life across the planet. At the same time, terrorists and other evildoers may develop a powerful capability to destroy that life."

## Recent Examples of "Contentious Research" in the Life Sciences

Biological weapons differ from other weapons systems in a number of important respects. They generally are based on naturally occurring pathogens that have coevolved along with their hosts to possess features such as high infectivity, ease of transmission, and virulence. As a corollary, however, the effects of naturally occurring pathogens are limited by the evolutionary advantage gained by not eliminating their hosts. Among the many implications of the anticipated progress in biotechnology is the presumption that it may be feasible to create novel biological agents that are far more predictable and dangerous than any of the naturally occurring pathogens that have been developed as biological weapons in the past.

It may be difficult to engineer a more successful pathogen than those already present in nature that have been perfected by evolution for their niche in life. However, application of the new genetic technologies makes the creation of "*designer diseases*" and pathogens with increased military utility more likely.

There have been several recent examples of what Gerald Epstein of the Defense Threat Reduction Agency refers to as "*contentious research*" — experiments that resulted in the creation of organisms or knowledge with "dual use" potential. The Australian *ectromelia* virus (mousepox) experiment; total synthesis of the poliovirus genome and recovery of infectious virus, and the comparison of the immune response to a host defense function from *vaccinia* and smallpox have all attracted the attention of the scientific community, the media, the defense community, and policy analysts. Each is elaborated below.

**Mousepox Virus: A Case Study in Preconsideration**

Probably the most celebrated recent case involving the dissemination of research with the potential for bioterrorist uses was the report of an unexpected effect of the bioengineering of a strain of *ectromelia* virus (mousepox) that was intended to help eradicate mice in Australia. The authors of the paper had originally set out to make an infectious immunocontraceptive for wild mice by incorporating a gene encoding an antigen from fertilized mouse eggs into the genome of *ectromelia* virus. Since the expression of this egg antigen by the virus did not result in infertility, the authors attempted to increase the virulence of *ectromelia* with the hope that this would increase the immune response of the infected mice to their fertilized eggs and thus make them infertile. They drew upon previous published work by others with recombinant *vaccinia* virus in mice in which it had been shown that incorporating the gene for the immunomodulatory cytokine IL-4 into the viral genome and thus overexpressing it *in vivo* enhanced the virulence of *vaccinia* virus in mice. The increased virulence is probably due to suppression of the antiviral immune response mediated through competing cytokines like IL-2, IL-12, and interferon gamma, which work by stimulating immune effector cells to kill virus-infected cells and thus control the virus infection.

The authors of this study used standard and quite simple procedures for incorporating the IL-4 gene into the mousepox genome. They then demonstrated that this engineered mousepox virus was more virulent than the parent virus and killed 60 percent of infected mice, even if the mice were from a genetically resistant strain. Even more unexpected was their observation that mice that had been vaccinated and were completely resistant to the parent virus, and even to a more virulent strain of mousepox, were now killed by the IL-4 gene-expressing virus.

Some have felt that the publication of this paper provides a blucprint or roadmap for terrorists to engineer a more virulent strain of smallpox that could overwhelm the human immune system in even well-vaccinated individuals. The methods section of the paper illustrates how easy it is to make an IL-4 expressing orthopox virus. It has been suggested that either the paper should not have been published, or at the very least the "*materials and methods*" section of the manuscript should have been altered or omitted entirely from the published article. The authors were sensitive to these issues and consulted with their peers in the Pest Animal Control Cooperative Research Center at the Australian National University in Canberra about whether the paper should be submitted for publication. Further, the editors of the *Journal of Virology* were sufficiently concerned about the implications of the research described in the article that they also sought the guidance of others about whether to publish the paper either intact or in some modified form. In the end the *Journal of Virology* decided to publish the manuscript as submitted.

This example illustrates the difficulty of attempting to censor either the initiation of research or the publication of results. The initial goals of the mousepox research were directed to control the population densities of a rodent pest in Australia. The studies were done on mice, and the virus itself, while related to smallpox, is not of any danger to humans. Thus, these studies had desirable scientific and societal goals and there was no obvious reason not to undertake them. Even in retrospect, the decision by informed scientists who had no vested interest in the work to approve publication seems appropriate. There were numerous examples in the published literature demonstrating the effects of cytokines, like IL-4 on immune modulation. The authors of this study, therefore, were building upon an established literature in this field that is filled with similar findings on the effects of the decreased or increased levels of IL-4 and other immunomodulatory factors on the virulence of other viruses and many microorganisms. As previously noted, the design of the mousepox study built upon previously published studies in which *vaccinia* virus engineered to express IL-4 was studied in mice. There is also a relevant preexisting literature in the field of oncology in which the increased expression of various cytokines incorporated into the fowlpox genomes and other orthopox viruses along with tumour antigens has been used to increase the immune response to tumours and decrease the immunogenicity of the viruses. The technique for incorporating new genes into the poxvirus genome had been published in many places. Thus there is little technical information that was not already abundantly available in the literature and well known to the scientific community.

The observation that even vaccinated mice were killed by the IL-4 expressing mousepox was a somewhat surprising finding that is of potential concern. However, since the ability of immunomodulatory factors to increase the virulence of this virus could have been predicted and the means to make such a virus were readily available, it was important to publicize that this strategy could overcome vaccination because it alerted the scientific community to such a possibility occurring either intentionally or spontaneously.

First, knowledge of these experiments allows the scientific community to explore how to overcome such engineered viruses. It informs us of the fact that we should monitor cytokine levels in the blood of the initial cases of a highly virulent virus that is used in an attack. Second, it suggests that we should be prepared to treat infections caused by such an engineered virus with antibodies that inactivate the relevant cytokine, with gamma interferon that would counter the effect of IL-4, or with both. Finally, it is worth noting that this work was done outside the United States and could have been published in an Australian or European journal, illustrating the limits of national policies to address dual use concerns and, in this case specifically, the need to have international guidelines for the publication of manuscripts containing "*sensitive*" information.

**Total Synthesis of the Poliovirus Genome and Recovery of Infectious Virus**

Wimmer and colleagues reported that they had reconstructed poliovirus from chemically synthesized oligonucleotides that were linked together and then transfected into cells. This report attracted considerable attention in the news media and concern in some segments of the public. The media treatment of the work suggested that this experiment proved that one could synthesize any virus from chemical reagents that can be purchased on the open market.

This implication raised the public concern about bioterrorism because it suggested that the Wimmer experiment provided a recipe for terrorists to manufacture the virus. In response to the publication of this article in *Science*, in the 107th Congress, Representative Dave Weldon (R-FL) introduced H. Res. 514, which criticized the publication of this research because of its implications for compromising the national security interests of the nation. The Weldon resolution, which did not pass, went on to state the concern of the House of Representatives regarding the potential of the poliovirus paper to enable terrorists to synthetically create a human pathogen to release on the people of this country and further called upon the publishers and editors of scientific publications and the scientific community to establish ethical standards and exercise restraint in the dissemination of information of potential use to terrorists in the development of bioterrorism agents. The Weldon resolution also called upon the Executive Branch to "examine all policies, including national security directives, relevant to the classification or publication of federally-funded research to ensure that, although the free exchange of information is encouraged, information that could be useful in the development of chemical, biological, or nuclear weapons is not made accessible to terrorists or countries of proliferation concern."

Many scientists concluded that the Wimmer experiment was neither a novel discovery nor a potential threat. The general principle that one could make live poliovirus from a DNA template was already known in 1981, when Baltimore and colleagues reported that a DNA copy of the positive strand RNA genome of poliovirus could be taken up into living cells under appropriate conditions and result in the generation of encapsulated, infectious virus. These studies led to the ability to manipulate DNA copies of RNA viral genomes to generate preselected genetic changes. This technology bypasses the technical problems of working with RNA molecules and allows subsequent recovery of infectious virus. Subsequent research has succeeded in extending this technology to RNA viruses with larger positive strand genomes, negative polarity RNA, or segmented genomes.

Several points should be emphasized, however. Like the mousepox IL-4 experiment discussed above, the technology for producing and manipulating the genome of RNA viruses has been available in the literature for a long time. The ability to synthesize a poliovirus genome and recover infectious virus was regarded as a foregone conclusion. The Wimmer approach offers no technical advantage to a terrorist. And more importantly, in fact it is a very labourious and difficult way to accomplish this synthesis. The interesting scientific results from the Wimmer experiment were not its highly touted potential for bioterrorism, but rather the fact that the virus synthesized had significantly weakened pathogenicity as compared to wild-type strains of poliovirus. The decreased virulence is likely due to third-base and noncoding changes inserted as supposedly neutral markers.

## Comparison of the Immune Response to a Virulence Gene from *Vaccinia* and Smallpox

*Variola major* virus causes smallpox, which has a 30-40 percent mortality rate, whereas *vaccinia* virus, which is used to vaccinate humans against smallpox, causes no disease in immunocompetent humans. In a paper that appeared in the *Proceedings of the National Academy of Sciences*, Rosengard and colleagues investigated a possible basis for the difference in the putative virulence factor between the virus that causes human disease and the one used to vaccinate against the disease. Both viruses have an inhibitor of immune response enzymes—*vaccinia* virus complement control protein (VCP) and smallpox inhibitor of complement enzymes (SPICE). The authors focused on a comparison of the genes encoding this inhibitor. As live *variola* is not available for study, they used standard techniques to synthesize the *variola* SPICE gene. They found that *variola* SPICE has a greater degree of specificity for human complement and is nearly a hundredfold more active than VCP at inactivating this component of the human immune system (human complement component C3b). The authors suggested that the difference between VCP and SPICE could explain the difference in virulence between the two viruses and the restriction of *variola*'s host range to humans.

Some might argue that the Rosengard study is of greater concern than the previous two examples because it provides information on how to increase the virulence of *vaccinia* virus, and thus on how to convert a readily available agent that has minimal virulence into a virulent virus. A commentary written on this paper pointed out that it is very unlikely that *vaccinia* virus carrying SPICE in place of VCP would approach the pathogenicity of *variola*. Furthermore, publication of the article alerted the community of scientists to this mechanism for virulence. This information should stimulate scientists in both the public and private sector to identify compounds or immunization procedures that disable SPICE. These could form the basis for new treatments or vaccines both to immunize against the naturally occurring smallpox virus and to counteract the genetically engineered variety.

## Response of the Life Sciences Community to Previous Challenges

As the preceding examples make abundantly clear, there is an increasing awareness within and outside the scientific community of the dangers posed by the proliferation of biological weapons capabilities. This heightened awareness has also increased the collective concerns of this Committee and the scientific community about preventing the destructive applications of biotechnology research. This is not a completely new issue. When gene splicing technology was first reported, the scientific community at the time raised concerns that the technology might deliberately or inadvertently be used to create organisms with increased virulence or novel characteristics. These possibilities eventually led to the 1975 Asilomar Conference, where scientists gathered to discuss the safety of manipulating DNA from different species. The meeting resulted in the issuance by NIH of Guidelines for Research Involving rDNA Molecules (hereafter called the NIH Guidelines) in 1976 that regulated the conduct of NIH-sponsored recombinant DNA research and established a mechanism for reviewing proposed experiments in this field.

Just as the life sciences community with the Asilomar Conference stepped up to the challenge of responding to concerns that biology could set back rather than advance human welfare, so too the Human Genome Project created the ethical, legal, and social implications program to explore how advances in genetics intended to improve human health could proceed without undermining other dimensions of human well-being. National commissions and Congress continue to debate whether certain advances in biology should be pursued and published.

The initial fears about the inadvertent creation of virulent microbes by gene splicing techniques have abated because of overwhelming scientific evidence to the contrary. There have been no reported cases of disease caused by recombinant microorganisms despite the widespread use of gene splicing techniques in academic laboratories and in the production of pharmaceuticals. In view of this experience, and the prospects for understanding the etiology of complex diseases and finding cures for them, the NIH has revised its Guidelines several times, with the net result being the elimination of the earlier prohibitions and the exemption from the Guidelines of essentially all recombinant DNA experiments except those that involve the molecular manipulation of human and restricted animal and plant pathogens.

## Committee Charge and Process

Current policy at both the national and international level may not be adequate to cope with the dangers inherent in the use and applications of genetic engineering. As discussed in greater detail in the following chapters, the United States has enacted legislation to provide for the physical security of select agents and screening of personnel. The Committee's proposed system for reviewing research projects and publications would complement and strengthen this statutory regime.

Internationally, however, protection against misuse of biotechnology is very uneven. The Biological and Toxin Weapons Convention, the centerpiece of biological weapons arms control, lacks effective verification and compliance measures. Moreover, it addresses only the actions of states and was never intended to guard against the development of a BW capability by individuals or nonstate actors (although national implementing legislation required by Article IV of the Convention could constrain the actions of individuals and groups within the state).

In November 2001 the draft text for an international protocol covering compliance and verification measures was rejected by the United States. New, informal measures to strengthen the BWC being explored by expert groups and States parties are scheduled to continue over a period of three years (the first meetings were held in Geneva in August 2003). The measures include enactment of national criminal legislation supplemented by an enhanced extradition regime; security standards for pathogenic organisms; genetic engineering oversight; and international adoption of professional codes of conduct. The hope is that these discussions will translate into coordinated action by the States parties, but at present only a few states have instituted security measures to protect against diversion and misuse of biotechnology.

The most elaborate treaty-based inspection procedures could not achieve effective restrictions at the level of basic research without severely restricting research in general. The inevitable diffusion of knowledge and capabilities has already demonstrated that the capacity to do harm is becoming globally available, both to state and nonstate actors. At the same time, developments in biotechnology are also capable of yielding great benefits, such as new treatments for many diseases.

The distinction between the great opportunities and great dangers of biotechnology turns on assessing whether the risk(s) associated with the benefits of fundamental research outweigh its potential for misuse. The challenge to the scientific community, therefore, is to develop formal and informal processes and procedures to mitigate or minimize the destructive applications of advanced biotechnology without unduly restricting legitimate biotechnology research activities.

Beginning the process of addressing these challenges is the purpose of this study. Specifically, the Committee was charged to:

1. Review the current rules, regulations, and institutional arrangements and processes in the United States that provide oversight of research on pathogens and potentially dangerous biotechnology research, within government laboratories, universities and other research institutions, and industry. The review would focus on how choices are made about which research is and is not appropriate, and how information about relevant ongoing research is collected and shared.
2. Use the review to assess the adequacy of current U.S. rules, regulations, and institutional arrangements and processes to prevent the destructive application of biotechnology research.
3. Recommend changes in those practices that could improve U.S. capacity to prevent the destructive application of biotechnology research while still enabling legitimate research to be conducted.

This report is part of a larger body of work that The National Academies have undertaken in recent decades on science and security issues, beginning with *Scientific Communication and National Security* in 1982 and continuing into the 1990's with the publication of *Chemical and Biological Terrorism: Research and Development to Improve Civilian Medical Response* (1999) and *Firepower in the Lab: Automation in the Fight Against Infectious Diseases and Bioterrorism* (2001). In response to the events of September 11th, the Academies undertook a comprehensive survey of the contributions that science and technology could make to countering terrorism; *Making the Nation Safer: The Role of Science and Technology in Countering Terrorism* was published in 2002.

The report of its panel on bioterrorism, *Countering Bioterrorism: The Role of Science and Technology*, was published separately. In addition, the Institute of Medicine's Forum on Emerging Infections convened a 3-day workshop on *Biological Threats and Terrorism: Assessing the Science and Response Capabilities*, which was released as a workshop summary late in 2002. The 2002 report on *Countering Agricultural Bioterrorism*, a study already in progress prior to September 11th, enabled the Committee to focus its primary efforts on threats to human health. In the area of potential controls on information and data, the report on *Sharing Publication – Related Data and Materials: Responsibility of Authorship in the Life Sciences* (2003) is particularly relevant to the continuing concerns for ensuring the wide availability of the results of scientific research. Other reports and information about current projects may be found on the Academies website.

**Committee Process**

In creating the Ad hoc Committee on Research Standards and Practices to Prevent the Destructive Application of Biotechnology, the National Research Council (the operating arm of The National Academies) selected committee members representing a broad spectrum of backgrounds, expertise, and interests. Areas of expertise included molecular and cellular biology, virology, medicine, laboratory safety, international and regulatory law, bioethics, and defense policy. In addition, the Committee relied on the expertise and advice of representatives from the Executive Office of the President, governmental and nongovernmental technical and policy experts, as well as educators and private consultants. Information available from the open literature and materials submitted by experts were reviewed and considered during the Committee's deliberations.

Even though the Statement of Task did not require the Committee to consider information control regimes for dissemination of information in the life sciences that could be exploited for nefarious purposes, the Committee concluded that this issue was implicit in the larger task before it and needed to be considered along with the regulatory environment for biotechnology research. An additional impetus for the Committee's consideration of information control regimes for unclassified research in the life sciences was the announcement by the White House shortly before the Committee's first meeting of its renewed interest in the application of "*sensitive but unclassified information*" control regimes for managing the dissemination of unclassified research that is financed by the federal government.

## Biological Warfare in History

People figured out how to intentionally spread illnesses long before naturalists came up with the discovery that germs cause disease. Among the older military techniques that can be claimed as biological warfare is the use of corpses of humans or animals to befoul wells or other sources of drinking water. While the principal objective was thought to be the denial of clean water to the enemy, a secondary effect was to spread disease among people and animals that consumed the contaminated water. One of the earliest recorded instances of biological warfare occurred in 600 BC, when the Athenian leader Solon poisoned the water supply in the city of Kirrha with the noxious roots of the *Helleborus* plant—a primitive but effective biological toxin of plant origin. The Greeks and Romans may have used human and animal corpses to poison drinking water wells. Alexander the Great is thought to have catapulted the bodies of dead men over the walls of besieged cities, possibly as a means of spreading disease and inciting terror among their inhabitants.

A related technique, used in the Middle Ages, was to deliberately leave dead human or animal corpses behind in areas that would be occupied shortly by invading troops; catapults were used as well. In 1346, invading Tartars intent on controlling the Silk Road trade attacked the Black Sea port of Caffa—at the time occupied by the Genoese. The Tartar army, already exposed to the Black Death, hurled plague-infested cadavers over the impregnable walls of Caffa to infect the enemy population. It is usually reported that the fleeing Genoese brought the Black Death with them—via plague-infested rodents, along shipping routes to Sicily, Sardinia, Corsica, and Genoa—and from there it spread overland throughout Italy and Europe. It is considered equally likely, however, that the entry of plague into Europe from the Crimea occurred independent of this event. Over a four-year period, the plague eventually caused 25 million deaths—one-third of Europe's population at the time. Population losses were probably much higher in the French Mediterranean coastlands and in northern Italy.

During the seventeenth and eighteenth centuries, French and British soldiers and civilians are alleged to have deliberately infected North American Indian populations with European diseases. "(T)he use of smallpox as a weapon may have been widely entertained by British military commanders and may have been employed without scruple when opportunity offered, possibly on a number of occasions." During the French and Indian Wars, for example, Sir Jeffrey Amherst, commander-in-chief of the British forces, was concerned that his troops west of the Allegheny Mountains were in danger of being overrun by Indians. He wrote to the commander of the garrison at Fort Pitt on the Pennsylvania frontier and urged that smallpox be spread among the disaffected tribes. In June 1763, Captain Ecuyer of the Royal Americans met with two Indian chiefs under a pretense of friendship and gave them blankets that had been taken from a smallpox hospital. During the following months, according to historians of the episode, many Indians suffered and died as "smallpox raged among the tribes of the Ohio." During the 1800s, U.S. government agents were alleged to have deliberately infected the Plains Indians by giving them trading blankets infected with the deadly disease, decimating the population.

# 11

# Regulatory Environment

The regulatory environment for the life sciences has been developed over the course of five decades. Responsibility for regulation in the United States of various aspects of biotechnology research in tne life sciences is shared among a number of federal agencies, ranging from the Department of Agriculture (USDA) and the Environmental Protection Agency (EPA) to the Nuclear Regulatory Commission (NRC). The National Institutes of Health (NIH), for example, sets standards and procedures for the research it funds on recombinant DNA (rDNA). Research on human gene therapy is a special case, with both NIH and the Food and Drug Administration (FDA) conducting reviews prior to the initiation of research. While review is mandatory for NIH-funded research, industry often seeks review voluntarily. The Centers for Disease Control and Prevention (CDC) sets standards for the handling and transport of some especially dangerous biological pathogens. The NRC has responsibility for regulations to control the receipt, possession, use, transfer, and disposal of radioactive materials by research institutions. The USDA has broad responsibility for biotechnology research related to plants and animals. Industries involved in biotechnology research generally have internal procedures to review potential research protocols, although these vary considerably within and between industrial and commercial facilities. Important aspects of their work such as clinical or field trials are also subject to regulation by federal agencies, in particular the FDA and the EPA. Universities have various methods for reviewing and approving potentially contentious research. Professional societies address questions of ethics and norms for research.

Until the mid-1990s the regulatory environment focused on protecting the public health and general environment from biological hazards associated with possible exposures to human pathogens via interstate transport, recombinant organisms, and containment of recombinants and their products so that inadvertent or deliberate releases of these materials to the environment would be within acceptable limits. The regulatory environment for microbial hazards also encompasses the importation of nonnative plant and animal pathogens.

Following the historic Asilomar Conference in 1975, the NIH, in 1976 published the Guidelines for Research Involving rDNA Molecules (hereinafter referred to as the NIH Guidelines or the Guidelines). The NIH Guidelines described four levels of combinations of laboratory practices, containment equipment, and facility safeguards that were thought to be appropriate for the safe use and physical containment of rDNA molecules in research. The four levels, P1 to P4, provide increasing levels of physical protection against personnel contact with or accidental release to the environment of genetically engineered microorganisms.

The CDC and the NIH encouraged the life sciences community to participate in a collaborative initiative to develop consensus guidelines to safeguard worker safety and public health from hazards associated with the possession and use of human pathogens in microbiological and biomedical laboratories. This initiative resulted in the publication by CDC and NIH in 1984 of Biosafety in Microbiological and Biomedical Laboratories (hereinafter referred to as the BMBL). These consensus guidelines also established four ascending levels of physical containment using the terminology Biosafety Level 1-4. The combinations of standard and special microbiological practices, safety equipment, and facilities for each level are similar to those of the NIH Guidelines. Specific recommendations for appropriate practices, equipment, and facility safeguards are given in the BMBL for pathogens that meet one or more of three criteria: the pathogen is a proven hazard to laboratory personnel working with infectious materials; the potential for laboratory-acquired infection is high even in the absence of previously documented laboratory-associated infections; or the consequences of infection are grave. The recommendations are advisory and are intended to provide a voluntary guide or code of practice for investigators who possess and use human pathogens in their research activities.

The Recombinant DNA Advisory Committee (RAC) and the BMBL process have been highly successful. Laboratory-acquired infections from exposure to biological agents known to cause disease are infrequent. There are no reports that the possession and use of biological agents and toxins for research, education, and other legitimate purposes endangers the public health. The fourth edition of the BMBL states: "Experience has demonstrated the prudence of the Biosafety Level 1-4 practices, procedures, and facilities described for manipulations of etiologic agents and laboratories settings and animal facilities. Although no national reporting system exists for reporting laboratory-associated infections, anecdotal information suggests that strict adherence to these guidelines does contribute to a healthier and safe work environment for laboratorians, their coworkers, and the surrounding community." This experience indicates that compliance with voluntary guidelines can achieve safety in research and clinical laboratories and protect the public health without significantly restricting the pursuit of science.

Spurred by rising concerns about bioterrorism, we are now witnessing a transition from an environment based upon voluntary compliance with recommended practices to a greater number of statutes and regulations, particularly for control of biological materials and personnel. It took the United States three years to ratify the 1972 Biological Weapons Convention (BWC) and 17 years for the United States Congress to pass legislation making the provisions of the BWC binding on all Americans. Not much changed until 1996 when, with the passage of the Antiterrorism and Effective Death Penalty Act, new regulatory controls were enacted swiftly regarding transfers of dangerous pathogens. Less than a year following the terrorist attacks on September 11, 2001 and subsequent anthrax mailings, two major pieces of legislation were passed by Congress and signed into law — "The Uniting and Strengthening America by Providing Appropriate Tools Required to Intercept and Obstruct Terrorism of October 2001" (hereinafter, the PATRIOT Act), and "The Public Health Security and Bioterrorism Preparedness and Response Act" of June 2002 (hereinafter, the Bioterrorism Response Act).

The PATRIOT Act makes it illegal in the United States for anyone to possess any biological agent, including any genetically engineered organism created by using rDNA technology, for any inappropriate reason. The Act also prohibits the transfer or possession of a listed biological agent or toxin by a "*restricted person.*" A "*restricted person*" is not permitted to ship or transport via interstate or foreign commerce, or possess, or receive any biological agent or toxin that has been shipped or transported in interstate or foreign commerce, if the biological agent or toxin is listed as a select agent.

The Bioterrorism Response Act added new requirements for the secretaries of the Departments of Agriculture and Health and Human Services to consider in listing agents and in preventing unlawful access to agents during transfers. The statute also establishes new requirements for registration with the appropriate secretary concerning possession and use of listed agents and toxins, including "information regarding the characterization of listed agents and toxins to facilitate their identification, including their source; and safeguard and security requirements for registered persons." The law also requires the Secretary to establish rules that provide appropriate physical security requirements for listed agents and for the Department of Justice — through the FBI—to conduct background investigations on individuals who are permitted access to select agents or who work in a facility where select agents are stored.

The security provisions in the Bioterrorism Response Act are radically transforming the life sciences research environment in the United States from one that is basically open to one that excludes, based upon criteria stipulated in the PATRIOT Act, certain individuals from access to and research on certain listed agents. The FBI provisions, which went into force without public notice and comment rulemaking, prescribe the collection of pertinent background information on individuals; who may access, use, receive or transfer select agents, and the release and disclosure of that information to other entities in the FBI Information Form. These provisions have raised concerns that qualified individuals may be discouraged from conducting biomedical and agricultural research of value to the United States because of the apparent infringement of these rules on individual liberties under the Fourth Amendment.

The next section of this chapter expands on the brief descriptions above to give a more complete picture of the current system of regulations and voluntary practices that govern research in biotechnology. It adds discussion of the growing web of controls over foreign nationals seeking to work, study, or participate in scientific activities in the United States and of the various codes of professional conduct that are a fundamental part of the self-governance of scientific practice. As noted at the beginning of this chapter, at present this system is focused on occupational safety and health and on environmental protection, but increasingly, additional efforts are being made to control access to biological materials that might be used by terrorists. With the exception of research involving human subjects, the system is not intended to provide oversight of research in the sense of making decisions about whether particular projects or experiments are appropriate.

## THE U.S. REGULATORY ENVIRONMENT

### Oversight of Genetic Engineering Research

The response of the life sciences community in the mid-1970s to concerns about the potential unknown risks inherent in research involving the new field of genetic engineering. The Asilomar process led to the NIH assuming responsibility for promoting safe conduct of such experiments and to the subsequent publication of the NIH Guidelines. The guidelines are designed to address the risks to public health and the environment associated with exposure to either rDNA molecules or organisms or viruses containing such materials. The NIH Guidelines are applicable to all rDNA research within or outside the United States or its territories where the research is conducted at an institution that receives any support for the research from the NIH, including research performed directly by NIH.

Institutions that are recipients of NIH support for rDNA research must establish an Institutional Biosafety Committee (IBC) as part of their compliance with the NIH Guidelines. Further, as part of documenting that they have established a properly constituted IBC, institutions must register the IBC with the NIH Office of Biotechnology Activities (OBA). IBCs are the cornerstone of institutional oversight of rDNA research.

An IBC is a review body appointed by an institution to review and approve potentially biohazardous lines of research relating primarily to rDNA research. IBCs were originally established to provide local, institutional oversight of nearly all forms of research utilizing rDNA. On behalf of the institution, IBCs review rDNA research projects for compliance with the NIH Guidelines. Over time, the role of the IBCs at many institutions has been expanded to include review and oversight of a variety of experimentation that involves biological materials (e.g., *infectious agents*) and other potentially hazardous agents (e.g., *carcinogens*).

While an IBC must consist of at least five members there is no upper limit on the number of members. Every IBC is required to have two members not affiliated with the institution who represent the interests of the surrounding community with respect to health and protection of the environment. These may be officials of state or local public health or environmental protection agencies, members of other local governmental bodies, or persons active in medical, occupational health, or environmental concerns in the community. It is also recommended that IBCs include: experts in biosafety and containment; persons knowledgeable in institutional policies and applicable laws; individuals reflecting community attitudes; and at least one representative member from the laboratory staff. Committee members cannot review a project in which they have been, or expect to be, involved or have a direct financial interest. Finally, the Guidelines provide that while opening IBC meetings to the public is suggested but not required, minutes of the meetings and submitted documents must be available to the public on request.

Because the NIH Guidelines require establishment of an IBC when research is conducted at or sponsored by an entity receiving any NIH support for rDNA research, even privately funded projects employing rDNA must adhere to the NIH Guidelines if they are being carried out at, or funded by, an organization that has any NIH contracts, grants, or other support for this kind of research. Additionally, some communities and real estate leases require compliance with the NIH Guidelines, making such compliance legally binding even for private companies. Adherence to the NIH Guidelines is mandatory and important because they stipulate biosafety and containment measures for rDNA research. Furthermore, they delineate critical ethical principles and outline key safety reporting requirements for human gene transfer research.

Most of the 400 or so IBCs registered with OBA are at institutions that are subject to the NIH Guidelines and for whom IBC registration is mandatory. While most of these institutions are academic, some industry-based IBCs are registered with NIH as a consequence of receiving NIH support for rDNA research (e.g., SBIR grants) and thereby becoming subjcct to the NIH Guidelines. In other instances, companies voluntarily comply with the NIH Guidelines as a means of observing the highest standards for safety practices; as part of that voluntary compliance, they register their IBCs with the NIH. Several federal agencies including the USDA and the Department of Veterans Affairs (VA) have made compliance with the NIH Guidelines a condition of their support of intramural and extramural research projects. Furthermore, a number of federal IBCs are registered with

NIH, including those at the NIH, the Department of Energy (DOE) laboratories (including the Lawrence Livermore, Los Alamos, Oak Ridge, Sandia National Laboratories) and various VA medical centers and military research institutes such as the Uniformed Services University of Health Sciences, the Walter Reed Army Medical Center, and the U.S. Army's Medical Research Institute for Infectious Diseases.

Some of these facilities are registered with NIH because they receive NIH support for their rDNA research and others because it is the policy of the department or agency to comply with the NIH Guidelines. The responsibility for the "*enforcement*" of the guidelines is shared by the NIH Office of Biotechnology Activities, the Recombinant DNA Advisory Committee (RAC), IBCs at individual institutions, and by the principal investigators (PIs) themselves.

## Framework for Implementation of the NIH Guidelines for rDNA Research

The Guidelines provide an administrative framework that specifies the roles and responsibilities of various federal officials, research institutions, and individual scientists. Significant responsibility is shared among the NIH Office of Biotechnology Activities (OBA), the RAC, IBCs at individual institutions, the principal investigator (PI), the Biological Safety Officer (BSO) at the institution, and by investigators themselves. Scientific advice on the technical aspects of risk assessment is provided by technical experts on the RAC; public input is provided by experts in nontechnical subjects and by the right of the public to comment on major actions.

The system is based upon a tiered set of reviews that encourages experimental design to be well thought out and provides a means for catching potential problems. The Guidelines distinguish among experiments: those needing approval of the IBC as well as the RAC and NIH director before initiation; those involving human testing that need approval of the IBC and the Institutional Review Boards (IRBs) as well as the RAC; those that require approval of the OBA and IBC; those that only require IBC approval; those that merely require notice to the IBC at the initiation of the experiment; and exempt experiments.

The RAC is designated to consist of up to 21 voting members, including the chair. A majority of the voting members have to be knowledgeable in relevant scientific fields, such as molecular genetics, molecular biology, or rDNA research, including clinical gene transfer research. At least four members of the RAC have to be knowledgeable in fields such as public health, laboratory safety, occupational health, protection of human subjects of research, the environment, ethics, law, public attitudes, or related fields. Representatives of various federal agencies also serve as nonvoting members. Over time, the degree of centralized federal oversight has been substantially reduced. Many of the central functions of the RAC have been delegated to IBCs. Each institution (and the IBC acting on its behalf) has become responsible for ensuring that all rDNA research conducted at or sponsored by that institution is conducted in compliance with the NIH Guidelines.

The RAC is, however, still responsible for advising the NIH director on actions such as: (1) adopting changes in the NIH Guidelines; (2) assigning containment levels, changing containment levels, and approving experiments considered "*Major Actions*" under the NIH Guidelines; (3) promulgating and amending lists of classes of rDNA molecules to be exempt from the Guidelines because they do not present a significant risk to health or the environment; and (4) certifying new host vector systems.

The RAC is also responsible for: (1) identifying novel human gene transfer experiments deserving of public discussion; (2) transmitting to the NIH director specific comments/recommendations about human gene transfer experiments; (3) publicly reviewing human gene transfer clinical trial data and relevant information evaluated and summarized by the NIH OBA in accordance with the annual data reporting requirements; (4) identifying broad scientific, safety, social, and ethical issues relevant to gene therapy research as potential Gene Therapy Policy Conference topics; (5) identifying novel social, ethical, scientific, and safety issues relevant to specific human applications of gene transfer and providing the necessary guidance.

As noted above, all institutions subject to the NIH Guidelines are required to establish and register an IBC for the review of rDNA research. The IBC is designed to provide a quasi-independent review of rDNA work done at an institution. It is responsible for: (1) reviewing all rDNA research conducted at or sponsored by the institution and approving those projects in conformity with the Guidelines; (2) periodically reviewing ongoing projects; (3) adopting emergency plans for spills and contamination; (4) lowering containment levels for certain rDNA and recombinant organisms in which the absence of harmful sequences has been established; and (5) reporting significant problems, violations, illnesses, or accidents to the NIH OBA.

It is also the responsibility of the institution to appoint a Biological Safety Officer if it engages in large-scale research or production activities involving viable organisms containing rDNA molecules. If the institution engages in rDNA research at BL-3 or BL-4 (see below), the officer must be a member of the IBC. The officer's duties include: (1) conducting periodic inspections to ensure laboratory standards are rigorously followed; (2) reporting to the IBC and the institution any significant problems, violations of the Guidelines, and any significant research-related accidents or illnesses; (3) developing emergency plans for handling accidental spills and personnel contamination and investigating laboratory accidents involving rDNA research; (4) providing advice on laboratory security; and (5) providing technical advice to the PI and the IBC on research safety procedures.

Pre-initiation review of experiments by the RAC has been an important part of the oversight mechanism. Pre-initiation approval of experiments by NIH is required only for: (1) experiments that have not been assigned containment levels by the Guidelines; (2) experiments using new host-vector systems, which must be certified by NIH; (3) certain experiments requiring case-by-case approval; and (4) requests for exceptions from Guideline requirements. Prior to the initiation of these experiments the PI must submit a registration document to the IBC containing the following information: the source(s) of DNA; the nature of the inserted DNA sequences; the host(s) and vector(s) to be used; whether an attempt will be made to obtain expression of a foreign gene, and if so, the protein that will be produced; and the containment conditions that will be implemented as specified in the NIH Guidelines.

The initial RAC review process includes a determination as to whether the human gene transfer experiment presents characteristics that warrant public RAC review and discussion. The NIH OBA will notify the PI(s) about the results of the RAC's initial review. Two outcomes are possible: (1) the experiment does not present characteristics that warrant further review and discussion and is therefore exempt from public RAC review and discussion; or (2) the experiment presents characteristics that warrant public RAC review and discussion. Completion of the RAC review process is defined as: (1) receipt by the PI(s) of a letter from the NIH OBA indicating that the submission does not present characteristics that warrant public RAC review and discussion; or (2) receipt by the PI(s) of a letter from the NIH OBA after public RAC review that summarizes the committee's key comments and recommendations (if any).

## Types of Experiments That Require IBC, RAC, and NIH Director Review

At this time, only two categories of experiments are considered "*major actions*" that require decision by the NIH director after review by the IBC and the RAC. One category includes experiments that propose the "deliberate transfer of a drug resistance trait to microorganisms that are not known to acquire the trait naturally—if such acquisition could compromise the use of the drug to control disease agents in humans, veterinary medicine, or agriculture." The second category includes experiments that propose the deliberate formation of rDNA-containing genes for the biosynthesis of toxin molecules lethal for vertebrates at an LD50 of less than 100 nanograms per kilogram of body weight (e.g., microbial toxins such as the botulinum toxins, tetanus toxin, diphtheria toxin, and *Shigella dysenteriae* neurotoxin). The containment conditions or stipulation requirements for such experiments must be recommended by the RAC and set by NIH at the time of approval.

## Physical and Biological Containment Strategies for NIH-funded rDNA Research Activities

Regulated experiments must be carried out in accordance with physical and biological containment levels; the degree of containment is based upon the degree of potential hazard. Physical containment requires practices, equipment, and facility safeguards that lessen the chances that a recombinant organism might escape. As discussed above, the NIH first published safety guidelines in 1976, followed by the publication in 1984 of *Biosafety in Microbiological and Biomedical Laboratories* (BMBL). The BMBL guidelines address laboratory safety procedures for working with and handling infectious disease agents—they do not address laboratory security issues.

The BMBL categorizes infectious agents and laboratory activities into four classes or levels (BL-1 to BL-4) and establishes safety requirements for each level based upon risk. Factors considered in determining the level of containment include agent factors such as: virulence, pathogenicity, infectious dose, environmental stability, route of spread, communicability, operations, quantity, availability of vaccine or treatment, and gene product effects such as toxicity, physiological activity, and allergenicity.

Experiments involving levels 2 through 4 and restricted risk group host organisms require IBC approval before recombinant experiments can be conducted. At the highest level (BL-4), nothing that is created should have any possibility of escape or of coming in direct contact with any laboratory workers. The containment conditions or stipulation requirements for such experiments must be recommended by the RAC and set by NIH at the time of approval.

Containment conditions for experiments involving the introduction of rDNA into restricted agents are set on a case-by-case basis following NIH OBA review. The recommended practices, safety equipment, and facility safeguards in these guidelines establish a code of practice that is complied with voluntarily, one that all members of a laboratory community can together embrace to safeguard their colleagues and to protect the public. A permit is also required for all facilities working with such agents, although clinical laboratories used for research, diagnostic, reference, and/or verification purposes need only be certified (but do not require a license).

Some organisms, including smallpox (*Variola major*) may not be studied in the United States except at specified facilities. Smallpox is an acute contagious disease caused by *Variola* virus, a member of the orthopox virus family. It was one of the world's most feared diseases until it was eradicated by a collaborative global vaccination program led by the World Health Organization (WHO). The last known natural case was in Somalia in 1977. Smallpox was officially declared eradicated in 1980. All research activities, including storage of *Variola major* are restricted to two international collaborating centers for smallpox research. The WHO Collaborating Center for Smallpox Research in the United States is located at the CDC in Atlanta, Georgia, the other is located at the VECTOR Laboratory in Koltsovo, Russia.

Since their initial appearance, the physical biocontainment levels for rDNA experiments have been progressively lowered over time. As experience provided confidence that rDNA technology could be applied without creating dangerous organisms that could not be contained, the prohibitions were replaced with a series of risk-based mechanisms for oversight and approval.

## Compliance with and Enforcement of NIH Guidelines

The PI is responsible for full compliance with the Guidelines in the conduct of rDNA research and for ensuring that the reporting requirements are fulfilled; the PI is held accountable for any reporting lapses. For experiments that require NIH approval prior to IBC approval, it is the responsibility of the PI to petition NIH OBA with the concurrence of the IBC.

Compliance with the Guidelines is accomplished by a combination of local selfregulation and limited federal oversight, with the ultimate enforcement resting in the federal funding power. Even if noncompliance were found, no penalties can be imposed other than restriction or termination of NIH funding. The primary mechanism in the Guidelines for enforcing compliance is local self-regulation. Noncompliance may result in: suspension, limitation, or termination of financial assistance for the noncompliant NIH-funded research project and of NIH funds for other rDNA research at the institution, or a requirement for prior NIH approval of any or all future rDNA projects at the institution.

The Guidelines are designed to encourage industry's voluntary compliance by creating a parallel system of project review and IBC approval analogous to that required for NIH-funded projects, modified to alleviate industry's concerns about protection of proprietary information. A company's IBC determines whether the facilities meet the standards for the large-scale containment level but only for information-gathering purposes rather than to enforce these guidelines assigned by the RAC. A working group of the RAC may visit the companies and their IBCs from time to time. An important provision here is a process whereby a corporation may request presubmission review of the records needed to register its projects with NIH. The HHS Freedom of Information Officer informally determines whether the records have to be released; if so, they are returned to the submitting company.

## Regulation of Microbial Agents (Listed Agents and Toxins)

The Antiterrorism and Effective Death Penalty Act of 1996, required the Secretary of HHS to establish and enforce safety procedures for the transfer of listed biological agents (select agents), including measures to ensure proper training and appropriate skills to handle such agents, and proper laboratory facilities to contain and dispose of such agents. These regulations provide:

1. "safeguards to prevent access to listed biological agents for use in domestic or international terrorism or for any other criminal purpose
2. procedures to protect public safety in the event of a transfer or potential transfer of a listed biological agent in violation of the established safety procedures and safeguards

3. the appropriate availability of biological agents for research, education, and other legitimate purposes."

The select agent list, which is subject to revision, includes those agents considered to be the greatest threats to human health. An expanded list of pathogens and toxins went into effect on February 11, 2003. Agricultural plant and animal pathogens are now also included; the other changes reflect taxonomic changes and a few reassessments of what constitutes the most dangerous biothreat agents.

The PATRIOT Act makes it a criminal offense for any person to knowingly possess any biological agent, toxin, or delivery system of a type or in a quantity that, under the circumstances, is not reasonably justified by prophylactic, protective, bona fide research, or other peaceful purpose. In addition, the new law prohibits transfer or possession of a listed biological agent or toxin by a "*restricted person.*"

Title II, Enhanced Controls of Dangerous Biological Agents and Toxins, of the Bioterrorism Response Act substantially broadens the regulatory obligations for laboratories working with select agents. The Secretary of HHS has the authority to establish and enforce safety procedures, including:

1. proper training and appropriate skills to handle such agents and toxins;
2. proper laboratory facilities to contain and dispose of such agents and toxins;
3. measures to prevent access to such agents and toxins for use in domestic or international terrorism or for any criminal purpose;
4. procedures to protect the public safety in the event of a violation of the safety or security measures; and
5. appropriate availability of biological agents and toxins for research, education, and other legitimate purposes.

On February 7, 2003 the CDC's final interim rule, Possession, Use and Transfer of Select Agents, went into effect. On February 11, 2003 similar rules from the Animal and Plant Health Inspection Service (APHIS) of the USDA also went into effect. A USDA permit is required for work with plant or animal pathogens. In accordance with accepted scientific and regulatory practices of the discipline of plant pathology, an exotic plant pathogen (e.g., virus, bacteria, or fungus) is one that is not known to occur within the United States. Determination of whether a pathogen has a potential for serious detrimental impact on managed (agricultural, forest, grassland) or natural ecosystems is made by the PI and the IBC, in consultation with scientists knowledgeable about plant diseases, crops, and ecosystems in the geographic area of the research.

These regulations impose additional shipping and handling requirements on laboratory facilities that transfer or receive select agents capable of causing substantial harm to human health. They are designed to ensure that select agents are not shipped to parties who are not equipped to handle them properly or who lack proper authorization for their requests. The major shift in the new regulations establishes who may possess select agents as well as who may send and receive those agents, adds biosecurity requirements to the biosafety requirements, incorporates the personnel restrictions of the PATRIOT Act, involves the FBI in performing background checks on individuals who may have access to or conduct research on select agents, and proscribes certain types of experiments.

## Possession of Select Agents

The Antiterrorism and Effective Death Penalty Act of 1996 required the Secretary of HHS to issue regulations governing the transfer of biological agents that have the potential to pose a severe threat to public health and safety. The CDC was authorized to regulate transfers of pathogens of unique interest in terms of their capacity to be used as weapons (the select agents list). Accordingly, the CDC required that laboratories transferring select agents be registered.

The Bioterrorism Response Act adds new requirements for the Secretary of HHS to consider in listing agents and in preventing unlawful access to agents during transfers. Facilities that register their possession and use of listed agents and toxins must provide "information regarding the characterization of listed agents and toxins to facilitate their identification, including their source; and safeguard and security requirements for registered persons."

Regulations specified under this law must "include appropriate safeguard and security requirements for persons possessing, using, or transferring a listed agent or toxin commensurate with the risk such agent or toxin poses to public health and safety (including the risk of use in domestic or international terrorism)." Registered facilities must limit access to listed biological agents and toxins only to those determined by the registered facility to have a legitimate need to handle or use select agents, and the Secretary must be notified if a listed agent is lost, stolen, or released outside a biocontainment area of a facility.

## Importation and Interstate Shipment of Etiologic Agents

The importation or subsequent receipt of human pathogens and vectors of human disease is controlled by the Public Health Service Foreign Quarantine Regulations. Packages containing human pathogens or vectors originating in foreign locations must have an importation permit issued by the CDC. The importer is legally responsible for ensuring that the foreign personnel package, label, and ship the infectious materials according to the Interstate Shipment of Etiological Agents regulations, regulations of the Department of Transportation on Transportation of Etiologic Agents, and the Dangerous Goods Regulations of the International Air Transport Association. An applicant for a permit must be knowledgeable of safe practices and proficient in the handling of infectious materials, be directly responsible for work with the infectious materials, and reside at the receipt address for the facility where work with the material will occur. The permit application requires the importer to provide characterization information for the material, a description of the objectives of the intended use, and a designation of the biosafety level of the laboratory where the work will occur.

The CDC is also responsible for regulating the interstate shipment of indigenous human pathogens, diagnostic specimens, and biologic products. The shipment of these materials must be in compliance with the provisions of the Interstate Shipment of Etiological Agents regulations, which specify packaging and labeling requirements and procedures for notification of successful delivery or failure of delivery.

## Oversight of Foreign Nationals

This section briefly describes the current and emerging system of granting permission, to non-U.S. citizens through the visa system, for both short-term and extended stays, as well as two of the tracking systems. The system is still evolving, so any description of current practice runs the risk of becoming rapidly out-of-date. At present, however, September 11th and its aftermath have significantly increased the level of scrutiny, the time involved, and the opacity of the process. It should also be noted that, beyond the requirements to designate responsible individuals in affected institutions, to date laws and regulations related to individuals have been almost entirely aimed at rejection and prevention. That is, they have been aimed to limit access rather than to create a process of licencing or certification that would convey some more general, authoritative approval for working in life sciences research comparable, for example, to the licenses doctors must obtain to practice medicine.

The September 11th terrorist attacks greatly increased the concern and accelerated the plans for improving efforts to provide adequate scrutiny of visa applications and to track foreign nationals once they entered the United States. Foreign scholars planning shorter visits are also affected by increased concern for security, with impacts on the ability of researchers to take part in international meetings, conferences, or international research collaborations. Over time, these various restrictions could potentially alter the way research is conducted and have the potential to impede scientific progress in the United States.

The Department of Homeland Security (DHS) has been given responsibility for the policy guidance and regulation governing the issuance of visas, with the Secretary for Homeland Security given ultimate authority to determine who may and who may not enter the United States. Where there are foreign policy considerations the Department of State will continue to exercise authority. Consular officers, who have responsibility for guiding the review and processing of visa applications, will also remain under the auspices of the Department of State. In testimony before the House Select Committee on Homeland Security on July 11, 2002, Secretary of State Colin L. Powell reported that the State Department adjudicated over 10 million nonimmigrant visa applications in 2001. Around 7.5 million visas, or about 70 percent of the total, were issued.

**Additional Security Checks on Visa Applications**

The Visas Condor program, initiated in January 2002, seeks to identify terrorists by checking a visa applicant's name against various U.S. government databases. Applicants are also required to fill out additional forms and be interviewed, fingerprinted, and subjected to additional identifying measures and background checks. Those affected by the Visas Condor program are predominantly Muslim men between the ages of 16 and 45 who come from any of approximately 26 (mostly Islamic) countries, but the system also applies to countries such as Russia and China. The State Department's goal is eventually to have the Visas Condor process take less than ten business days.

In response to earlier concerns the State Department, in consultation with other federal agencies, had created a Technology Alert List to provide guidance about which areas of science and technology were of particular concern. Applications from individuals with expertise in one of these areas would be sent to Washington for further review, usually by an agency with expertise in that field and perhaps by the FBI or intelligence services. The 16 categories on the list include "chemical and biotechnology engineering," which covers "technologies associated with the development or production of biological and toxin agents, pathogenics, biological weapons research." In practice, "*technologies*" tended to be defined broadly enough to affect life scientists doing a variety of research.

Since January 2002 the Visas Condor security checks and the Technology Alert List reviews have required explicit approval from Washington for each applicant. In the past, at least the Alert List review process permitted consular officers to issue visas if they had not received a negative report from Washington within a certain number of days, but that is no longer the case. The agencies that need to provide clearance are determined by the State Department's Bureau of Consular Affairs and include the CIA and the FBI, as well as any other agency with a potential interest in the applicant. All applicants must be positively cleared by all the agencies involved in the review. This led to the backlogs and time delays reported in recent months. Consular Affairs officers have reported that in 2002, they conducted 35,000 Visas Condor and about 14,000 other checks. While this represents about a threefold increase in the number of cases referred to Washington, D.C., it is nonetheless a very small percentage of the total number of cases.

In May 2003, Secretary of State Powell announced additional requirements for those seeking nonimmigrant visas. Except for certain visa categories or for countries where a visa waiver is in effect, as of August 1, 2003 all individuals between the ages of 16 and 60 are required to undergo a personal interview as part of the visa application process. Substantial delays and increasing backlogs are anticipated in the visa process, since no additional resources are being allocated to consular officers and no overtime is to be used to handle the additional interviews. Furthermore, a new legislative mandate also requires that, as of January 1, 2004, all visitors entering the United States on a visa will be photographed and fingerprinted as part of U.S.-VISIT, the enhanced security screening process.

**Tracking Systems**

In addition to increased scrutiny of visa applications, the U.S. government is initiating a number of systems for tracking foreign students and visitors to the United States.

### *Student and exchange visitor information system (SEVIS)*

The new Student and Exchange Visitor Information System (SEVIS) is an electronic System aimed at keeping better track of foreign students once they have received visas to study in the United States. The Bureau of Citizenship and Immigration Services (formerly the Immigration and Naturalization Service, which was incorporated into the Department of Homeland Security) is responsible for SEVIS, although the program was developed in cooperation with the Departments of State and Education. SEVIS is designed to collect and report data on international student or exchange visitor status and changes, such as a change in one's program of study. It also provides system alerts, event notifications, and basic reports to the end-user schools, programs, and INS field offices. The timetable for its implementation and for colleges and universities to come into compliance with its regulations was accelerated after September 11th. Schools wishing to accept foreign students were required to register with SEVIS by January 30, 2003.

### *Interagency panel for advanced science and security (IPASS)*

IPASS is a response to an October 2001 Presidential Decision Directive, "Combating Terrorism through Immigration Policies," which directed federal agencies to develop student immigration policies through which the country "prohibits certain students from receiving education and training in sensitive areas." The White House's Office of Science and Technology Policy (OSTP) has been working with the White House's Homeland Security Council and others to develop and implement IPASS, although at the time of this report the Executive Order to create IPASS had not yet been signed.

## Professional Education and Responsibilities of Life Scientists

The Center for the Study of Ethics in the Professions lists over 850 "*codes of ethics*" on its website. A code of ethics is that profession's contract with the society it serves establishing in often very general terms acceptable moral behaviour for practitioners of that profession. Some differ widely in their content, because of their origins and their specific purposes. Others are similar in the topics they cover and the general ethical standards they articulate, but differ in language and in the specific ethical problems or abuses they address.

There is a considerable literature on the formulation of professional oaths and codes of conduct. Some have called for the initiation of a pledge to be taken by scientists—perhaps at graduation—much as modernized versions of the Hippocratic oath are taken by some medical students upon graduation. Others focus less on the development of codes and more on the inclusion of an emphasis on the moral and social responsibilities of life scientists in the training of students and postdoctoral fellows. Particularly if efforts to address the social responsibilities of scientists are led by leaders in the field and senior investigators, it is argued, young scientists will come to value "the ethics of individual behaviour within the scientific enterprise and the societal impact of scientific research." Whether mandatory or voluntary, the adoption of codes of conduct by professional organizations or national academies of science, and the integration of ethics education into the training of students should serve to sensitize "young scientists to reflect on the wider consequences of their intended field of work."

Arguably, to be effective, any policy or set of procedures intended to address concerns about the offensive application of life sciences research data will require "*ownership*" by the scientific community. To the extent that responsibilities to guard against intentional misuse are recognized in professional codes of conduct and explicated and examined in the context of the training of the next generation of practitioners in the life sciences, opportunities to develop and maintain ownership by the community will only be increased.

At the November 2001 Review Conference for the BWC, the United States formally proposed new ways to strengthen the regime against biological weapons. Among the recommendations put forward was one that called upon the countries that are parties to the BWC to support the development and adoption of a code of conduct for scientists working with pathogenic organisms. Among the guiding principles of such a code of conduct would be a statement that "scientists will use their knowledge and skills for the advancement of human welfare and will not conduct any activities directed toward the use of microorganisms or toxins for hostile purposes or in armed conflict." This proposal will be taken up by an intersessional meeting of the parties to the BWC in 2005 in Geneva. Proposals for the creation of such professional codes of conduct for practitioners in the life sciences have also come from the International Committee of the Red Cross, the U.K. Foreign and Commonwealth Office, the European Union, The Royal Society, and others. Other U.S.-proposed measures to strengthen national and international implementation of the BWC include the oversight of "*high-risk*" genetic engineering experiments.

## International Situation

There is a deep and long-standing foundation of scientific self-regulation, voluntary standards, and associational accreditation. Given the fundamentally international character of research in the life sciences, any serious attempt to prevent the misuse of research must include efforts at improving and harmonizing standards and practices internationally. Recently, this has been supplemented by some mandatory requirements on specific aspects of laboratory safety.

This section provides a brief overview of some of the major international programs. It also offers examples from the regulatory systems of two other countries with advanced biotechnology research capabilities: the United Kingdom and Japan. The Committee's charge did not extend to a comprehensive review of the international regulatory environment, but the Committee did examine some of the existing systems for possible positive or negative examples that might be relevant to the evolving U.S. situation.

### Laboratory Safety

The Organization for Economic Cooperation and Development (OECD) has drafted quality management requirements called the Good Laboratory Practice (GLP) guidelines. Many national governments require laboratories that carry out safety and toxicological testing for the approval of new products to meet the GLP guidelines.

#### *Controls over access to and transfers of dangerous pathogens*

As discussed above, smallpox was declared to be eradicated by the WHO at the annual meeting of the World Health Assembly in May 1980. This led to the greatest international control over access to dangerous pathogens—an international agreement implemented by the WHO to restrict the repository of smallpox virus cultures to two designated facilities, one in the United States and one in Russia. All countries other than the United States and Russia were to destroy their remaining stocks. The WHO, however, had no enforcement authority or means of verification and relied entirely on the voluntary cooperation of member states, leaving uncertainties about compliance.

There are several other nonlegally binding access control agreements. The Australia Group is an informal arrangement of 33 member countries plus the European Union that harmonizes national controls on the export of dual use materials and production equipment that, in the wrong hands, could increase the risk of assisting chemical and biological weapons (CBW) proliferation. The Group meets annually to discuss ways in which each national-level export licensing measures can more effectively ensure that would-be "*proliferators*" are unable to obtain necessary inputs for CBW programs. Participants in the Australia Group do not undertake any legally binding obligations. By enhancing the effectiveness of national export licensing measures, the Australia Group's activities serve to support the objectives and purposes of the BWC. The participants in the Australia Group encourage all countries to take the necessary steps to ensure that they and their industries are not contributing to the spread of biological and chemical weapons. Export licensing measures demonstrate the determination of Australia Group countries to avoid involvement in the proliferation of these weapons in violation of international law and norms. The effectiveness of the cooperation among the participating countries depends solely on their commitment to CBW nonproliferation goals and the effectiveness of the measures they each take on a national basis.

The European Union also imposes export controls on dual use biotechnology equipment and pathogenic microorganisms and toxins, including agents that could be used for biological warfare. To date, it is the only regional organization to undertake such an effort.

International regulations apply far more comprehensively to transnational shipment of human, plant, and animal pathogens. Among the international organizations that set regulations controlling the international transfer of such material is the International Air Transport Association (IATA). IATA Dangerous Goods Regulations require that packaging used for the transport of materials in specified hazard groups meet defined standards. Shippers of microorganisms within the more serious hazard groups must be trained by IATA certified and approved instructors. They also require shippers' declaration forms, which should accompany the package in duplicate, and specified labels are used for organisms in transit by air (IATA, 1998).

The Universal Postal Union (UPU) has established strict regulations on the shipment of pathogens through the mail. Other organizations regulate specific modes of transport. These regulations are primarily directed to the prevention of accidental release, but they also operate to track (but not to limit) who is supplying and receiving pathogens. The European Union is the only regional organization to regulate the shipment of pathogens.

***Situation in the United Kingdom***

In light of the September 11th terrorist attacks in the United States, the focus has shifted from safety requirements in the laboratory toward greater scrutiny of dangerous substances and increasing the difficulty in gaining access to areas where such agents are stored and used. The Antiterrorism, Crime and Security Act 2001 (ATCSA) part VII, is instrumental in this approach and attempts to tighten controls on access to dangerous pathogens and toxins used in research establishments and laboratories in the U.K. The pathogens and toxins affected are specified in Schedule 5 of the Act classified with their ACDP hazard group. "*Dangerous substance*" means: (a) anything that consists of or includes a substance for the time being mentioned in Schedule 5; or (b) anything that is infected with or otherwise carries any such substance. Further substances may be added to the list by order of the Secretary of State.

In addition, the ATCSA establishes the power to vet personnel working in such establishments and to mandate security provisions. The owner of any premises that possess or use a dangerous substance must notify the Secretary of State. The occupier of the premises must ensure that only appropriate individuals are given access to the premises. A police officer may require provision of information about each person who has access to any dangerous substance kept or used there or who has access to specified premises and identify the access that the person has, or is proposed to have.

Moreover, a constable may require provision of information about what dangerous substances are kept or used at the premises, the measures taken to ensure the security of any such substance, and measures taken to ensure that access to the substance is given only to those whose activities require access and only in circumstances that ensure the security of the substance. A constable may require that measures be taken to ensure the security of any dangerous substance. To assess compliance with those measures, a constable may, after giving at least 2 days notice, enter any relevant premises at a reasonable time. A constable who has entered any premises may search the premises, building, or site; require any person who appears to the constable to be in charge of the premises, building, or site to facilitate any such inspection; and require any such person to answer any question.

If research establishments do not meet personnel or security requirements, access to dangerous pathogens and toxins could be withdrawn. Where the Secretary of State reasonably believes that adequate measures to ensure the security of any dangerous substance kept or used in any relevant premises are not being taken and are unlikely to be taken, he may give a direction to the occupier of the premises requiring him to dispose of the substance. Moreover, the Secretary of State may give directions to the occupier of any relevant premises requiring him to secure that the person identified in the directions is not to have access to any dangerous substance kept or used there nor to specified premises. The Secretary of State may not give the directions unless he believes that they are necessary in the interests of national security. Failure to comply with the relevant duties is punishable by imprisonment, a fine, or both.

## Research Oversight

### *United Kingdom*

For research involving DNA, the U.K. has set up the Health and Safety Executive (HSE) under the Health and Safety at Work Act of 1974 (HWSA). It is primarily concerned with the protection of human health from possible ill effects of any workplace activity. Genetic modification and any activities in which genetically modified cells or organisms are cultured, stored, used, transported, destroyed or disposed of, under conditions of containment, are subject to the control of HSE under the Genetically Modified Organisms (Contained Use) Regulations of 1992, which are made pursuant to the HSWA.

The following bodies were established specifically to provide policy guidance on issues arising from developments in modern biotechnology. The Advisory Committee on Genetic Modification (ACGM) is a nonstatutory body that advises the Health and Safety Commission/ Executive and Ministers on human and environmental safety of the contained use of genetically modified organisms under the GMO (Contained Use) Regulations 1992 (based on EC Directive 90/219), as amended. ACGM focuses on safety questions in the laboratory and industrial installations. It is not involved in policy approval. The Advisory Group on Scientific Advances in Genetics (AGSAG) is a nonstatutory advisory body that advises the Chief Medical Officer and the Director of Research and Development (DH) on potential implications for public health and for the National Health Service (NHS) of scientific advances in genetics. It also advises the NHS executive board on innovative genetic services and their evaluation.

### *Japan*

Guidelines for rDNA experimentation define basic conditions required to promote and ensure safety for rDNA and related experiments. The experiments must be conducted under proper safety measures generally employed in microbiological laboratories, incorporating combinations of physical and biological containment measures as required by the safety evaluation of the experiment. Large-scale experiments with genetically engineered organisms must be conducted in a facility that has appropriate containment measures. Laboratory workers must be aware of the necessity of safety measures in the experiments, actually take those measures, and must have been thoroughly trained to ensure their expertise in standard methods and practices in microbiological experiments.

## Conclusions

International regulation of biology is complicated by the lack of a multilateral consensus as to the basic security framework to which controls can be consistently applied. In contrast, the International Atomic Energy Agency (IAEA) oversees the Nuclear Nonproliferation Treaty (NPT) and nuclear safeguards agreements are negotiated with member states on a bilateral basis. The Organization for the Prohibition of Chemical Weapons oversees implementation of the exceptionally detailed Chemical Weapons Convention. Nothing comparable exists with regard to the oversight of biolotechnology. The BWC articulates a widely shared global norm against the weaponization of pathogens and establishes statutory but not regulatory obligations on parties to the Convention. Nor is there any international oversight organization for biology. Efforts to strengthen the BWC by adding provisions for verification and compliance foundered in 2001 on fundamental diplomatic differences of principle and in particular cost-benefit analyses as to the effectiveness of the measures being proposed and whether multilateral versus bilateral approaches were the best way to prevent the development of biological weapons.

Multilateral discussions are continuing on ways to strengthen compliance with this treaty. At the recent BWC review conference in November 2002, member states agreed on a U.S. proposal to hold intersessional meetings in each of the next three years (2003 - 2005) before the 2006 Review Conference to discuss the five voluntary measures put forward by the U.S. to strengthen the BWC.

With regard to oversight of research, *no country* has developed guidelines and practices to address all aspects of biotechnology research. There are a range of norms, standards of conduct for research, regulations, and institutional practices, many of which have been developed to address questions about research involving human subjects or the treatment of laboratory animals. In addition, responsibility for regulation of various aspects of biotechnology research is frequently shared among different departments or agencies.

In the United States, the PATRIOT Act and the Bioterrorism Response Act already establish the statutory and regulatory basis for protecting biological materials from inadvertent misuse. Once fully implemented, the mandated registration for possession of select agents, designation of restricted individuals who may not possess select agents, and a regulatory system for the physical security of the most dangerous pathogens within the United States will provide a useful accounting of domestic laboratories engaged in legitimate research and some reduction in the risk of pathogens acquired from designated facilities falling into the hands of terrorists. The Committee stresses that implementation of current legislation must not be overly restrictive given the critical role that the development of effective vaccines, diagnostics, therapeutics, and detection systems, along with a responsive public health system, will play in providing protection against bioterrorism—and other serious health threats. Otherwise these legislative solutions may unintentionally limit the research on dangerous pathogens by legitimate laboratories and investigators. To be effective, a harmonized, international system for the regulatory oversight of the possession of dangerous pathogens and toxins, comparable to the one being put in place in the United States, is needed.

Moreover, the different regulations now on the books do not add up to a systematic, generally applicable, means for the United States to respond to the challenges posed by research in the life sciences employing advanced biotechnology methods. Nor do they address the issues surrounding how to "*manage*" the knowledge and technologies produced through these research activities. At the moment, "*control*" over the results of these "*dual use*" research activities may be implemented at the point of information dissemination in the peer-reviewed literature. A critical question is whether the various regulations and laws can be adapted, enhanced, supplemented, and linked to provide a system of oversight that will give confidence that the potential risks of misuse of dual use research are being adequately addressed.

# 12

# SENSITIVE INFORMATION

The terrorist attacks of September 11, 2001 and the subsequent anthrax mailings in which five people died have produced a new sense of vulnerability in the United States. The governmental response has been wide ranging, affecting almost every sector of society. With respect to the life sciences, the most important initiatives to date are those embodied in the PATRIOT Act and the Bioterrorism Response Act. The latter legislation provides for the regulation of access to select agents and toxins through registration and screening of all institutions and individuals that possess, use, or transfer select agents.

Some have proposed that government control should go beyond the registration of the laboratories and researchers who work with specified agents to include broad controls on the dissemination of research results, as well as the vetting of research proposals. Should these proposals be adopted they would require a regulatory framework that would involve procedures for reviewing research proposals and restrictions on dissemination of research results; inevitably these regulations would profoundly affect research practices in biology laboratories. In effect, areas of life sciences research that were deemed "*sensitive*" because they could theoretically aid terrorists or be used in the production of biological weapons would be treated as secret.

Such a step should not be taken lightly; openness in science is highly valued. As the 1982 "Corson" report stated: Free communication among scientists is viewed as an essential factor in scientific advance. Such communication enables critical new findings or new theories to be readily and systematically subjected to the scrutiny of others and thereby verified or debunked. Moreover, because science is a cumulative activity—each scientist builds on the work of others—the free availability of information both provides the foundations for further scientific advance and prevents needlessly redundant work. Such communications also serve to stimulate creativity, both because scientists compete keenly for the respect of their peers by attempting to be first in publishing the answers to difficult problems and because communication can inspire new lines of investigation. Finally, free communication helps to build the necessary willingness to confront any idea, no matter how eccentric, and to assess it on its merits.

This chapter reviews the existing and emerging regulatory and oversight structure that governs the control of information related to biological research. Because issues of secrecy and sensitive information are new for much of the biological sciences, the chapter first discusses the experience of other scientific disciplines with these concerns. How other disciplines have addressed concerns about security suggests lessons that the Committee believes are relevant to the biological sciences as they respond to these issues.

## Past as Prologue?

The life sciences differ from the physical sciences in that they have not been deeply involved in developing new weapons in the United States since the Biological and Toxin Weapons Convention banned biological weapons in the early 1970s. While many countries pursued BW work prior to the BWC entering into force, only a few had large-scale programs, and even in those countries military support for biological research was dwarfed by the resources going into nuclear and conventional weapons programs.

The main patrons of research in the life sciences in the United States have been the National Institutes of Health (NIH), the Department of Agriculture, the National Science Foundation, the Department of Defense, the Department of Energy, and the pharmaceutical and agricultural industries. Secrecy issues involving national security concerns, as distinct from questions of intellectual property, have largely been absent in the life sciences. When the military did finance research in fields that had a biological content, such as oceanography, the result was to shift the balance in the field away from biology and toward the physical sciences that were familiar to the military program officers and advisory boards. It should also be recognized, however, that the Defense Department has had a long-standing interest in fundamental basic and applied medical research for the development of diagnostic and medical countermeasures to "*exotic*" diseases that could adversely affect personnel readiness.

## Nuclear Weapons Complex

It is instructive to compare the situation in the life sciences to other areas of science in which the military has taken a stronger interest. The U.S. nuclear weapons program offers an example in which the Departments of Defense and Energy have played dominant roles in funding and shaping developments in nuclear physics and related fields. Nuclear weapons design is carried out in the Energy Department's national laboratories (primarily the Los Alamos National Laboratory, Sandia National Laboratories, and Lawrence Livermore National Laboratory). This work requires a special security clearance that restricts access to a small number of scientists, who, in effect, constitute a closed society.

Although the principles that underlie the design of nuclear warheads are well understood by scientists around the world, the details of nuclear weapons design remain highly classified. Under the Atomic Energy Act of 1954, all information concerning nuclear weapons is "*born classified*," so that even research done outside the national laboratories under private sponsorship may be automatically classified if it is deemed relevant to nuclear weapons. The category of "*unclassified controlled nuclear information*" (UCNI) is exempt from release under the Freedom of Information Act. Exports of nuclear materials and related technologies are controlled under provisions of the Atomic Energy Act, the Nuclear Nonproliferation Act of 1978, and the Nuclear Nonproliferation Treaty, with nuclear dual use items covered by the Export Administration Act. As discussed below, "*technology*" can include information and various kinds of technical data and knowledge.

In short, there is a pervasive system of governmental secrecy and control for all research and development information related to nuclear weapons design and testing. Moreover, there is substantial consensus among scientists and the public that secrecy in the case of nuclear weapons is justified and should be maintained. Nuclear weapons scientists exchange their freedom to publish for relatively secure jobs and the satisfaction of feeling that they are contributing to national security.

Other structural elements further distinguish nuclear weapons from biological and chemical weapons. Although some of the knowledge and facilities related to nuclear energy are relevant to the production of nuclear weapons, many steps intervene between the underlying science and successful production of a reliable nuclear weapon. Production of weapons-grade material, for example, requires industrial-scale processes.

The proliferation of nuclear weapons capabilities has in every case so far required the resources of a state-sponsored program. This means that in the case of nuclear weapons there are relatively few interests that might argue for greater openness for research results. By contrast, in the case of biological weapons, the basic science relevant for civilian uses is essentially the same as that relevant to military—and especially terrorist—applications. No "*bright line*" exists between purely defensive and purely offensive uses of infrastructure and knowledge. Additional comparisons of the distinct differences between fissile materials and biological pathogens that fundamentally affect the security concerns related to research relevant to nuclear and biological weapons.

Some scientists have argued that a small group of terrorists, using knowledge that has long been publicly available, could assemble a crude bomb based on highly enriched uranium (HEU). Such a device would not be a weapon of the sort designed in state weapons programs; it could, nevertheless, potentially equal the explosive power of the bomb used at Hiroshima. This possibility makes the safeguarding of stockpiles of fissile materials paramount, since controlling access to HEU remains the primary technical barrier to this type of weapons proliferation.

**Cryptography**

Cryptography offers a second comparative case with potential lessons for the regulation of biotechnology. For hundreds of years cryptography was the province of governments that wanted to conceal state secrets—diplomatic and military—from others. To that end, they also kept information secret about the codes they used in order to make decryption by others less likely. The situation changed when private corporations developed a serious interest in cryptography as they began to do business electronically and needed to be able to conduct their affairs in private. Over the last three decades researchers in academic and industrial laboratories have entered the field in increasing numbers. Cryptography has thus become a dual use technology.

The new users of cryptography do not always see eye to eye with the government, which has an interest in retaining the ability to crack codes used by criminals and by foreigners doing business in the United States. In the 1990s controversy flared over the government's attempt as part of a new encryption standard to impose a key escrow feature, which would have enabled it to read any message it desired. The cryptography community and commercial interests argued that such a system would compromise privacy and undermine consumer confidence in business conducted over the Internet. It might also provide an opportunity for abuse of power by government agents holding the key. The encryption escrow feature was retained in the new standard, but use of the standard is voluntary.

More significant for the debate over secrecy was the government's attempt in the late 1980s to put some academic research papers dealing with cryptography under export control laws and to require that research publications be vetted by the National Security Agency (NSA). The security concern was (and is) that open publication of some research could reveal vulnerabilities in encryption algorithms that an enemy could exploit. The cryptography community, by contrast, has argued consistently that in a well-designed cryptographic system only the key should be secret. In addition, algorithms should be public and open to challenge, so that problems can be quickly identified and fixed; that is, cryptography, like other sciences, progresses best under conditions of openness.

The outcome of this controversy provides an alternative model for addressing security concerns. In this case, the government dropped its efforts to impose secrecy in exchange for an informal system in which cryptographers often (but not always) submit their proposals and papers to the NSA for prepublication review, even when NSA is not the funding agency. Thus, although there is no "*born secret*" category for cryptography research, the government has been able to keep track of ongoing research and to exercise some control over publication of results. Compliance is not universal, but there appears to be an informal norm in large segments of the cryptography community that cooperating with the government on this issue is a sign of good citizenship.

**Lessons from the Comparisons**

What lessons do these examples offer for possible governmental controls on research in the life sciences? We can compare structural conditions in the three cases in at least four dimensions. In all three cases, the government has an interest in controlling access to information for security reasons: there is a prima facie argument for keeping some research secret from potential enemies. The ease with which research results can be transformed into weapons or a technological advantage to be used against the United States varies sharply, however. In the case of nuclear weapons, understanding the principles behind the bomb is only part of what is needed to produce a weapon. Just as important —and far more difficult to obtain—is access to plutonium or weapons-grade uranium and the know-how to construct the device. Cryptography lies at the opposite extreme: a cryptographer can pursue his or her research with no more than pencil and paper, or at least with computing capabilities that are widely available.

Research in biology lies between these two extremes. Traditionally, biology has been considered a small-scale science. Although work in genomics, proteomics, and bionanotechnology is overturning this paradigm, the research and development associated with biological weapons programs do not necessarily require large-scale investment or specialized, dedicated facilities. Creating pathogen weapons poses certain technical challenges, but the ability to produce enough material to cause morbidity, mortality, public panic, and economic costs is within the capability of many laboratories.

The degree to which the three technologies are dual use also varies. The civilian uses of nuclear energy have been cordoned off from weapons developments through a large investment in security classification, international diplomacy, and a discourse that insists that nuclear weapons are special. Cryptography, as we have seen, has recently become a dual use technology, but its applications in the civilian world are growing rapidly in tandem with the Internet. The life sciences lie beyond cryptography in the dual use dimension; civilian uses dominate in the field, and the military's interest in the life sciences, although not negligible, has been dwarfed by their interest in the physical sciences. This is seen most clearly, perhaps, in the dominance until fairly recently of the NIH, NSF, and USDA, rather than DOD and DOE, in federal support of research in the life sciences.

The size of the field is also important when contemplating government controls. One reason the cryptography solution has worked relatively well may be that the number of publications is so small. The leading journal, *The Journal of Cryptology,* publishes only about 20 papers a year, and researchers present about 125 papers annually at conferences. By contrast, the American Society for Microbiology's 11 journals publish 6,000 papers a year, and by some estimates there are between 10,000 and 20,000 journals published in the life sciences internationally. Even if only a very small fraction of the research in these journals potentially arouses concern, the sheer volume of publication in the life sciences would make any effort to devise a screening mechanism for information deemed "*sensitive*" —or to ensure compliance—a daunting challenge.

In addition, there are many more life scientists than there are nuclear physicists or cryptographers. Only a few scientists are likely to be working with the list of select agents that have been the target of control so far, but many more would be included in a control regime that encompassed sensitive techniques as well as an expanded list of select agents. Furthermore, unlike the nuclear physicists, the life scientists are in many widely dispersed locations. The more people and facilities subject to control, the higher the costs.

Finally, there is no established culture of working with the national security community among life scientists as currently exists in the fields of nuclear physics and cryptography. As a group, biologists lack the experience of either nuclear physicists or cryptographers in interacting with the security agencies of the federal government, and conversely those agencies lack close ties and working relationships with the life sciences community.

The tradition of classified government research is well established in the latter two fields; the counterpart in the life sciences was the DoD program for research on biological weapons centered at Fort Detrick, MD, which ended in 1970. As noted above, however, that bioweapons program was only a small part of the government's funding of basic research in the life sciences and its very secrecy tended to isolate it from the larger community of life scientists. Since 1970, when President Nixon ended offensive biological weapons research, Fort Detrick has been a relatively open facility, housing a number of military and civilian tenants, including a large array of National Cancer Institute laboratories, as well as the U.S. Army Medical Research Institute for Infectious Diseases (USAMRIID).

The number of life scientists involved in intramural research sponsored by the Department of the Army remains, however, relatively small. It should be noted that other DoD agencies, such as the Defense Advanced Research Projects Agency (DARPA) and the Defense Threat Reduction Agency, have become significant sponsors of biodefense research, as have other federal agencies such as the Department of Homeland Security and the intelligence community. DARPA-sponsored research, while sometimes controversial, is unclassified.

The differences among the three areas are instructive. They suggest that controls on information flows in the life sciences will face obstacles rather different from those encountered in nuclear science and cryptography. The situation would be further complicated by the expansion of categories of information that the government wishes to control.

## Secret and Sensitive Information

An excellent summary of the different types of information control regimes in the United States is published by the Association of American Universities.

### Secret Information

The U.S. government handles issues of secrecy through a complex mix of statutes, regulations, and procedures that govern the control of classified information, public access to government information, and the maintenance of government records. Only those with security clearances are given access to classified information. Having a clearance, however, is not enough to provide access to classified information. An individual must also have a "*need to know*" the information in question. There is a formal system for controlling access to certain areas of information on a need-to-know basis. This additional layer of categories and controls adds to the complexity of the system.

With two exceptions, only one of which is relevant to the life sciences, designating information as secret requires an affirmative action by a government official and can be applied only to information created within an agreed framework that makes classification a possibility. That is, the government has no authority to designate information produced outside this framework as classified; in effect, the classification system applies only to work done in government laboratories or under government contract. The first exception, described above, is the Atomic Energy Act, where information related to nuclear weapons may be "*born classified*" without any prior involvement of the government in its generation.

The second exception, which is potentially relevant to aspects of biotechnology research, permits information received as part of the patent application process to be classified. Under the Invention Secrecy Act of 1951, the government is required to impose "*secrecy orders*" on certain patent applications that contain sensitive information. The disclosure of the invention is restricted and the grant of a patent is withheld. As summarized by the Project on Government Secrecy of the Federation of American Scientists, "[t]his requirement can be imposed even when the application is generated and entirely owned by a private individual or company. There are several types of secrecy order which range in severity from simple prohibitions on export (but allowing other disclosure for legitimate business purposes) to a classification requiring secure storage of the application and prohibition of all disclosure."

In the wake of the September 11th attacks, President Bush has extended classification authority to a number of agencies not previously involved in these matters, including the USDA, DHHS, and EPA. Their new authority will be exercised under the existing classification system.

Under the current system, in most agencies the task of managing potentially classified authority is so large that the authority to classify information has been delegated extensively. Literally thousands of government officials have classification authority. Detailed guides attempt to provide standards by which to judge whether particular information should be considered classified, but the more removed the issues are from specific details or facts, the more judgment becomes involved. One would expect the same delegation to occur as the USDA, DHHS and EPA exercise their new authority.

As discussed above, the struggle to decide whether areas of scientific research should be restricted in the name of national security recurred throughout the Cold War. During the early 1980s the Reagan Administration sought to restrict scientific communication in a number of fields. That controversy eventually led to a presidential directive in 1985, influenced in part by the Corson report. National Security Decision Directive 189 (NSDD-189) states that federally funded fundamental research, such as that conducted in universities and laboratories, should "to the maximum extent possible" be unrestricted. Where restriction is deemed necessary, the control mechanism is formal classification. "No restrictions may be placed upon the conduct or reporting of federally-funded fundamental research that has not received national security classification, except as provided in applicable U.S. statutes." This policy is still in force and was reaffirmed as recently as November 2001, pending completion of an administration review of export policy controls.

For all the serious concerns that can arise over whether information is properly classified, and how such decisions are made, debates over the classification of scientific research take place within a system of reasonably well-specified and understood rules. For the life sciences, with the exception noted above of the potential imposition of secrecy in the patent process, the question of whether research would be carried out under the restrictions of classification and whether research results might be classified would be part of the initial process of defining the terms under which the research is funded. Far more problematic is the interest in designating certain areas of research and certain types of knowledge in the life sciences—wherever they are produced and however they are funded—as "*sensitive but unclassified.*"

**Sensitive Information**

The issue of "*sensitive information*" is not new. Classification is only one of the ways in which the U.S. government controls public access to information. Across the federal government there are many other categories that apply narrowly or broadly to specific types of information. Some of the categories are defined in statute, some through regulation, and some only through administrative practices. Different agencies may also assign a variety of civil and even criminal penalties for violation of their restrictions.

The most extensive restrictions are exceptions to the Freedom of Information Act (FOIA), which enable the government to deny public access to particular classes of information. The withheld information, considered "For Official Use Only," must be tied to a particular FOIA exemption, for example, to protect individual privacy or proprietary business information. Some of the categories of exemptions are sufficiently general, however, to give federal agencies considerable latitude in withholding information related to internal decision-making.

At a time of heightened concern about the proliferation of weapons of mass destruction to states or terrorists, many kinds of information can seem potentially relevant to U.S. adversaries and hence targets for expanded controls. For scientists the chief concern in any government-imposed requirement to shield "*sensitive*" information lies in the potential fuzziness of the category, coupled with the severity of possible penalties for failing to protect the information. The most relevant categories of restricted information for research in the biological sciences are those related to "*sensitive but unclassified*" information and to "*dual use*" information covered by export controls.

**Defining "Sensitive" Information**

Sensitive but unclassified (SBU) information includes information generated within the government but may also extend to knowledge generated purely in the private sector. Particularly in the wake of the September 11th attacks, the standard examples of sensitive information tend to be those that relate to the vulnerability of critical infrastructures, including facilities that are privately owned. The key nodes in an electricity grid are a frequently cited example, and information related to the design and operation of a nuclear power plant or transport of nuclear materials has long been protected as Unclassified Controlled

Nuclear Information. In addition, it is easy to imagine that information such as the location of biological research programs might be considered sensitive, if the theft of select agents is considered a threat. The Bioterrorism Response Act exempts information on possession of select agents from FOIA.

The Bush Administration has urged federal agencies to use all applicable exemptions to the Freedom of Information Act when considering requests for "*sensitive but unclassified information*." The White House assigned the Office of Management and Budget the task of developing uniform policy guidance for government agencies on defining and controlling sensitive information. Section 892 of the Homeland Security Act directs the President to "identify and safeguard homeland security information that is sensitive but unclassified." No definition of "*sensitive*" is provided in the statute, however, and the fact that different agencies have put forward different definitions is a further concern. A key question is whether restraints on sensitive information might be extended beyond the information held internally by federal agencies and, if so, who would be responsible for determining what counts as "*sensitive*."

The DOE provides an illustrative example of the difficulties associated with attempts to define "*sensitive*" information. In the wake of the scandals over alleged Chinese spying at DOE laboratories in 1999, the Defense Authorization Act for FY2000 added a provision that imposed signficant civil penalties for disclosure of "*sensitive*" information even though no implementing regulations were ever produced. In January 2000 the DOE General Counsel took the position that, since no definition of sensitive information existed in the Atomic Energy Act or departmental regulations, legal restrictions could not be applied or enforced on DOE employees or federal contractors.

More generally, one basic DOE document defined "*sensitive but unclassified information*" as:

Information for which disclosure, misuse, alteration, or destruction could adversely affect national security or government interests. National security interests are those unclassified matters that relate to the national defense or foreign relations of the Federal Government. Government interests are those related, but not limited to, the wide range of government or government-derived economic, human, financial, industrial, agricultural, technological, and law enforcement information, as well as the privacy or confidentiality of information provided to the Federal Government by its citizens.

DOE also maintains a "*Sensitive Subjects List*" to guide what information can be discussed with non-U.S. citizens without triggering the need for an export license. One list promulgated in 1999 included a set of topics under "*Chemical and Biological Weapons*" described as "*illustrative but not exhaustive*." Information regarding these topics "may pertain to the research, design, development, testing, manufacturing, production, or use." The proposed topics included:

1. Genetic research, techniques, and specialized equipment related to chemical or biological agents, *for example*:
   (a) genome sequences and databases;
   (b) genetic engineering techniques
2. Defenses against, and vulnerabilities to, the use of chemical or biological agents, *for example*:
   (a) vaccines, antitoxins;
   (b) equipment, including protective clothing

This example is directly relevant only to DOE or DOE-contractor employees—although important work on the human genome is conducted at DOE facilities. Nonetheless, it illustrates the dilemmas scientists could face if there were a concerted government effort to promulgate and enforce such categories.

Recently, however, DOE has sought to clarify its policies, in part to address concerns about whether real or anticipated restrictions were hampering the vitality of the national laboratories. The Department is attempting to implement a policy in which "*Official Use Only*" information, based on the Freedom of Information Act, will be the standard for deciding whether and how to control unclassified information, gradually replacing the various other information categories that have emerged over the years. In addition, a department-wide memo on May 12, 2003 reaffirmed that the provisions of NSDD-89 remain the basis for DOE policy regarding restrictions on fundamental research. These DOE policies would presumably need to be reconciled with the regulations that will be required to define and implement Section 892 of the Homeland Security Act, which created the new category of "*sensitive homeland security information*."

**Dual Use Information**

Scientists may also face restrictions on their communications with foreign colleagues under various export control restrictions on sharing information regarding dual use technology. These restrictions can apply to communication both within the United States and with scientists abroad. As with the Technology Alert List designed to prompt scrutiny of visa applications, the export controls governed by the Export Administration Act and its implementing regulations also extend to the transfer of dual use technology. Technology is considered "specific information necessary for the '*development*,' '*production*,' or '*use*' of a product," and providing such information to a foreign national within the United States may be considered a "*deemed export*" whose transfer requires an export license. Technology "which arises during or as a result of fundamental research" is not subject to export restrictions—which relieves many scientists—but not those engaged in proprietary research sponsored by commercial interests at public and private universities.

## Publication of Sensitive Information in the Life Sciences

Until recently, there were very few cases of problems related to the publication of research results in the life sciences that attracted significant public attention. Some specialists in bioterrorism, however, had warned that, given continuing advances in biotechnology, open publication could provide information of use to terrorists. The publication of the "*mousepox*" study, made the issue a major concern for journal editors. The public perception of potential risks associated with publication of such information led to calls for scientific journals to refrain from publishing "*dangerous*" research or to delete some data from published research results in order to preclude others from replicating the results. Journals in the life sciences have responded in a number of ways to the concerns that published articles might provide useful knowledge or a roadmap for terrorists or rogue states.

In addition to the results of fundamental research, the compilation, synthesis, and assessment of already published results in review articles may provide an understanding of a field that could guide or assist terrorists. Even more difficult are the concerns raised by reports that result when scientists are assembled to render their judgment as experts about particular problems, even when they rely completely on open sources of information. Against these risks, one must weigh the genuine service to the research community provided by review articles and the contributions of expert panels to informed public debate and decision-making on issues where scientific knowledge and perspective play a role. The Committee wanted to acknowledge these problems, which it expects will remain and perhaps grow as a concern, but they are beyond the scope of this report.

In response to the concerns about publication of research results, the American Society for Microbiology (ASM) determined that the 11 journals it publishes would not restrict the information in the materials and methods section of articles. But ASM has also instituted formal procedures as part of the peer-review process for submitted articles so that reviewers address the potential risks of the research results to national security. At present, these policies apply primarily—although not exclusively—to research conducted on select agents. In 2002, of the 13,929 manuscripts submitted to ASM journals, 313 select agent manuscripts received special screening, and of these two manuscripts received additional screening by the full ASM publication board. The statistics through July 2003 are 8,557 manuscripts submitted, 262 select agent manuscripts screened, and none referred to the publication board for further review. Other journals, such as *Science*, the *Proceedings of the National Academy of Sciences*, and *Nature*, have also become alert to potential articles that could cause concern and have moved to develop review procedures of their own.

These new procedures have been the subject of intense discussion within the life sciences community and between life scientists and the national security community. In January 2003, for example, the National Academy of Sciences and the Center for Strategic and International Studies convened a one-day workshop to review the general question of sensitive information and specific issues of publication. Gatherings such as this attest to the seriousness with which the scientific and national security communities regard these issues but also to the difficulty of establishing productive communication—and even more of devising satisfactory, workable solutions.

In mid-February 2003, the editors of the major journals in the life sciences, including *Nature*, *Cell*, *Science*, and the *Proceedings of the National Academy of Sciences* (*PNAS*), published a joint statement on "Scientific Publication and Security." The statement was the outcome of discussions begun at the January workshop. It has generated substantial comment and controversy, but is also an example of the efforts of the scientific community to respond to issues related to potential risks through the development of self-governance mechanisms.

## Conclusions

Any argument about imposing information controls—whether through formal classification or restrictions on "*sensitive*" information—must be made in the context of the specific institutional history and research culture of the life sciences research community. Like all sciences, the life sciences rely upon a culture of openness in research, where the free exchange of ideas allows researchers to build on the results of others, while simultaneously opening scientific results to critical scrutiny so that mistakes can be recognized and corrected sooner rather than later. Most scientists would argue that the openness that characterizes much of the scientific research enterprise is the source of the extraordinary gains in scientific knowledge that have enriched us materially and intellectually.

It is not that individual researchers, research groups, university administrators, or editors do not know how to keep secrets. Anyone who has spent much time in a university recognizes that there are categories of information that are not widely shared, from faculty salaries at private universities to the location of animal testing facilities. Academic journals and funding agencies keep secret the names of their reviewers. Research performed under contract to proprietary interests routinely requires a period of secrecy and prepublication review of manuscripts intended for presentation and publication in the peer-reviewed literature as a contractual condition of funding. These areas of secrecy, however, are the result of widely accepted understandings and local negotiations, and the number of people affected is limited to those directly concerned.

As already discussed, compared with other disciplines such as physics, the life sciences have relatively little experience with classified research. Beyond this, the life sciences cover a broad set of disciplines, from evolution and ecology to genomics and proteomics. Unlike nuclear weapons research, much of life sciences research is not of interest either to "*rogue*" offensive weapons programs or to potential terrorists. The range of scientists and institutions affected would thus be hard to enumerate, let alone monitor.

The costs of complying with information controls on life sciences research would range from their impact on the culture of the research laboratories, which is generally acknowledged to be extraordinarily open, to financial costs borne by institutions in complying with government regulations, to the creation of obstacles to monitoring compliance with international arms control measures directed at biological weapons. The restrictions already in effect on select agents have caused some laboratories to destroy archived samples and to limit exchanges of materials between scientists. To extend government controls to the information contained in laboratory reports, conference papers, and journal articles would further constrict avenues of communication, both formal and informal, which have been an essential source of the dynamism of biological research in the modern era.

Perhaps most important, major universities have proscribed classified research on campus. Those who do accept classified research have usually created separate facilities where access can be limited and controlled. Secrecy would thus deprive the government of the graduate students and postdoctoral fellows who drive much of biological research—in many cases the best minds engaged in rapidly developing fields. Even without formal classification, the specter of information controls on "*sensitive*" information, given the current vagueness of the categories and the great difficulty of being any more exact about most of the dual use research, could be a significant deterrent to scientists to undertake research in some areas, such as infectious diseases. Yet these are precisely the areas where the best researchers are needed to help develop the nation's defenses against biological weapons, bioterrorism, and emerging-disease threats.

Thus there is a danger that the life sciences as a field of study would come to be regarded as less inviting, affecting the quality of researchers entering the field or making it more attractive to work outside the United States. Unlike the situation with nuclear weapons sign/development/production and testing, biotechnology-related research in the life sciences is an international activity and proliferation-relevant knowledge is widely held. Limiting the development of biotechnology in the United States would reduce our worldwide competitiveness in this rapidly changing field. We conclude that imposing mandatory information controls on research in the life sciences, if attempted, will be difficult and expensive with little likely gain in genuine security. The next chapter describes the system that the Committee has concluded can best meet the needs of reducing the risks of misuse of biological research while still enabling vitally needed research to meet civilian and biodefense needs to go forward.

# 13

# RECOMMENDATIONS

The preceding chapters have reviewed the nature of the threat associated with "*dual use*" knowledge in the life sciences, the current regulatory environment for the conduct and reporting of genetic engineering research in the life sciences, both domestically and internationally, and various information control regimes developed over the last 60 years in the United States. The Committee has concluded that existing domestic and international guidelines and regulations for the conduct of basic or applied genetic engineering research may ensure the physical safety of laboratory workers and the surrounding environment from contact with, or exposure to, pathogenic agents or "*novel*" organisms. However, they do not currently address the potential for misuse of the tools, technology, or knowledge base of this research enterprise for offensive military or terrorist purposes. In addition, no national or international review body currently has the legal authority or self-governance responsibility to evaluate a proposed research activity prior to its conduct to determine whether the risks associated with the proposed research, and its potential for misuse, outweigh its potential benefits.

After extensive deliberation, the Committee recognized the importance of educating the biotechnology research community about the potential dangers posed by dual use of new technologies. Rather than considering methods to identify and prohibit certain areas of research, we believe the community should work together with government agencies to develop communication channels so that both are aware of potential problems. The Committee has concluded that a system is needed to build these channels of communication and to provide greater oversight for the research enterprise. The significant increases in funding that will be going to research on biodefense—precisely the sort of research likely to pose the most severe dual use dilemmas—reinforce the argument for creating such a system. As the case studies demonstrate, even experiments that have the greatest potential for diversion to offensive applications or terrorist purposes may also have potentially beneficial uses for public health promotion and defense. To proscribe such experiments without a thorough assessment of their potential risks and benefits carries the possibility for hindering our ability to detect, identify, and defend against the new threat environment.

The system we are proposing would establish a number of stages at which experiments and eventually their results could be reviewed to provide reassurance that advances in biotechnology with potential applications for *bioterrorism* or *biological weapons* receive responsible oversight. The system relies heavily upon voluntary self-governance by the scientific community and expansion of an existing regulatory process that itself grew out of an earlier response by the scientific community to the perceived risks associated with gene-splicing research.

The heart of the system would be a set of guidelines to help identify research that could raise concerns because of its potential for diversion to offensive military applications. The concept behind these guidelines is that they will provide criteria that can assist knowledgeable scientists, editorial boards of scientific journals, and funding agencies in weighing the potential for offensive applications against the expected benefits of an experiment in this arena. It is important to realize, however, that identifying these concerns will not prevent a determined nonstate actor or individual from doing harm. Moreover, the Committee was adamant that these concerns should not be interpreted as defining a category of "*sensitive but unclassified*" research. Rather, like the NIH Guidelines for Research Involving rDNA Molecules established in the 1970s to guide research in a then-new and possibly risky technology, they can serve as the basis for a continuing dialogue between the members of the scientific community, the national security community, and the public. And, like the rDNA Guidelines, they must be applied on a case-by-case basis, with the opportunity for revision as new knowledge and experience in their operation accumulate.

## Key Assumptions

In developing the system outlined in this chapter, the Committee based its recommendations on several key assumptions. Each is discussed in turn.

## Many More Experiments will Explore the Virulence Factors of Bacteria and Viruses

The great majority of experiments on pathogenic bacteria or viruses are performed to ascertain exactly what makes the microbes pathogenic and virulent. Scientists are thus continuously exploring the ways that turning certain genes "on" and "off" enable these agents to be transmissible or cause disease in an appropriate host organism. Moreover, the concern over bioterrorism has stimulated the government to provide significantly increased funding to help combat infectious disease. The Fiscal Year 2003 budget passed by Congress and signed into law by President Bush in January 2003 added \$6 - 10 billion, spread across a number of agencies, to biodefense research in the United States. The NIH, for example, received \$1.5 billion for biodefense research. Internationally, other countries are also increasing their investments in civilian bioterrorism defense research. These increased domestic and international investments in basic and applied public health and bioterrorism defense research will inevitably create an increased number of research activities that raise concerns about misuse. This increased activity will also undoubtedly increase the number of research practitioners in this ever-expanding field of investigation with a corresponding increase in the number of articles appearing in the peer-reviewed literature.

## Scientific Evaluation of the Risks is Essential

As we described earlier and provided brief assessments of the dual use dilemmas presented by three recently published experiments. Although quite different, each of these cases has generated significant controversy. In the judgment of many scientists, the publication of the synthetic poliovirus paper presented no great contribution to the field of virology and no obvious advantages to a bioterrorist. The risks for potential misuse generated by the information in the other two papers are certainly greater. This is not to say they should not have been published. Rather, the Committee believes these cases illustrate that, to balance these risks against the obvious benefits, one must depend upon expert scientific judgment.

In fact, the paper describing the engineering of the mousepox was judged both by local scientific officials in Australia and by the editorial board of the *Journal of Virology* to have scientific merit and, on balance, to provide important information required for the progress in fighting disease. The third paper stimulated an accompanying commentary by the journal when it was published. The commentary concluded that the benefits of the original research contribution to the understanding of the complement system far outweighed the risks that the information could be "*misused.*" But making such judgments requires scientific training and knowledge— and expertise in one field may not always provide sufficient understanding of the relevance of research results in another.

The qualitative and case-by-case nature of these judgments is a primary reason the committee believes it is better to rely on selfgovernance to manage this aspect of the problem rather than to attempt to define appropriate or inappropriate research via regulation. As discussed above, key aspects of the system we propose are intended to augment the existing statutory and regulatory framework for controlling biological materials and personnel through voluntary arrangements addressing research issues.

### Only an International Set of Standards will Help to Minimize the Misuse of Biotechnology

Although the focus of this report is on the United States, this country is only one of many pursuing biotechnology research at the highest level. The techniques, reagents, and information that could be used for offensive applications are readily available and accessible. And the expertise and know how to use or misuse them is distributed across the globe. Without international consensus and consistent guidelines for overseeing research in advanced biotechnology, limitations on certain types of research in the United States would only impede the progress of biomedical research here and undermine our own national interests. It is entirely appropriate for the United States to develop a system to provide oversight of research activities domestically, but the effort will ultimately afford little protection if it is not adopted internationally. This is a challenge for governments, international organizations, and the entire international scientific community. Efforts to meet that challenge are under way, but they must be quickly expanded, strengthened, and harmonized.

## Recommendations

The system we propose for the United States consists of a number of filters for research proposals and publication of results that would cumulatively serve to protect against potential misuse yet enable important research activities to go forward. The key initial filter is awareness in the research community of categories of research that should raise concerns and collective community commitment to actively manage such research. Voluntary restraint based on awareness should be supplemented by review through existing bodies, namely an Institutional Biological Safety Committee/Recombinant DNA Advisory Committee process augmented to include the assessment of the potential for misuse as a criterion for approval or denial of proposed experiments. At the stage of publication, we recommend enhancing and expanding the process begun by the editors of a number of the leading scientific journals in February 2003. Finally, since this new system relies on both regulatory and voluntary elements, and involves issues and relationships with which the life sciences community has little experience compared to its colleagues in other fields, we recommend creation of a National Science Advisory Board for Biodefense (NSABB) to provide advice and assessments to the government and the scientific community as the system of review we are proposing develops. We recognize that successfully implementing the system we propose will require significant additional resources at each stage; we do not attempt to provide an estimate of those costs. Otherwise, concerns for unfunded mandates could be a significant barrier to full consideration of the proposals by the scientific community.

In making its recommendations, the Committee has sought to strike a balance and propose processes and mechanisms that will raise awareness and alarms when needed, without unduly constraining the practice, processes, and products of the life sciences research enterprise. We believe that such a system in the United States could also serve as a model for similar restraint in other countries.

**Educating the Scientific Community**

We recommend that national and international professional societies and related organizations and institutions create programs to educate scientists about the nature of the dual use dilemma in biotechnology and their responsibilities to mitigate its risks.

Adequately addressing the potential risks that research in advanced biotechnology could be misused by hostile parties will require educating the community of life scientists, both about the nature of these risks and about the responsibilities of scientists to address and manage them. At present, awareness of the potential for misuse of biological knowledge varies widely in the research community. Researchers currently working with select agents are already taking steps to contain these agents physically and protect against planned or unplanned harm. But most life scientists have had little direct experience with the issues of biological weapons and bioterrorism since the advent of the Biological Weapons Convention in the early 1970s, so these researchers lack the experience and historical precedent of considering the potential for misuse of their discoveries.

Fortunately, an extensive national and international network of professional societies provides the natural basis for increasing knowledge and awareness about the potential risks of research in advanced biotechnology. These societies hold numerous professional meetings to share the results of research and address issues of concern to the research community. We recommend that the societies undertake a regular series of meetings and symposia at these gatherings, in the United States and overseas, to provide both knowledge and opportunities for discussion. It could be useful for one of the major professional societies or science policy organizations to convene a meeting of all the major societies to discuss how best to implement such a program. Industry groups and associations of higher education and research could also usefully undertake the education of their members about the risks and their implications for research practices.

Substantive knowledge of the potential risks is not sufficient, however. The Committee believes that biological scientists have an affirmative moral duty to avoid contributing to the advancement of biowarfare or bioterrorism. Individuals are never morally obligated to do the impossible, and so scientists cannot be expected to *ensure* that knowledge they generate will never assist in advancing biowarfare or bioterrorism. However, scientists can and should take reasonable steps to minimize this possibility. The Committee believes that it is the responsibility of the research community, including scientific societies and organizations, to define what these reasonable steps entail and to provide scientists with the education, skills, and support they need to honor these steps.

These principles should be added to the codes of ethics of relevant professional societies. Most scientists are familiar with and carefully respect the moral norms of their profession that focus on the pursuit of truth and the advancement of science. Often placed under the heading of research integrity, prohibitions against fraud and plagiarism, as well as affirmative duties of openness in the sharing of findings, are well understood. Concerns about potential conflicts of interest and respect for intellectual property are similarly well appreciated, if not as clearly delineated. The addition of the moral duty we endorse here to those more familiar to the scientific community is not as novel as it may appear at first. As some scientific societies have recognized, scientists also have a general moral duty to use their knowledge and skill for the advancement of human welfare. We are only providing a specification of that general moral responsibility.

We believe further that scientists have an obligation to inculcate these moral duties in the next generation, both by example and by specific education and evaluation of their trainees. Other models of training in social responsibilities need to be explored, for example from the law and from medicine. In the law, most students sit for a multistate professional responsibility exam. In medicine, many specialty boards now examine young physicians in ethics as well as in medical skill and knowledge. Scientists will need assistance in learning about these other models, but they need to take charge of how best to educate their own next generation.

Scientists also should be willing to assist efforts to integrate the advancement of knowledge with the protection of national security by volunteering their time to sit on relevant peer review committees and national bodies, much as scientists contribute to advancing science currently by serving on study sections and as reviewers for professional journals. Finally, if scientists are to embrace the moral responsibilities outlined here, their home institutions must provide accommodation and support. Service in review of protocols and in student training must not only be encouraged but also rewarded.

## Review of Plans for Experiments

We recommend that the Department of Health and Human Services (DHHS) augment the already established system for review of experiments involving recombinant DNA conducted by the National Institutes of Health to create a review system for seven classes of experiments (the Experiments of Concern) involving microbial agents that raise concerns about their potential for misuse.

This part of the system includes both the criteria for deciding which experiments will be subject to review and the process by which the review will take place.

### *Criteria for review*

The experience with rDNA experiments emphasizes the importance of guidelines developed by the scientific community itself. The guidelines for work with rDNA promulgated after the Asilomar conference in 1975 have proven remarkably flexible and effective. The logical structure of such a system, keyed to the probability of harm associated with exposure(s) to genetically modified organisms—which changed as risk perceptions change—has been integral to their success.

The guidelines have prevented any untoward events and have allowed the rapid and efficient progress of the academic and commercial applications of these technologies. We now need to build upon the Asilomar experience to develop a uniform set of criteria to manage this new set of risks.

The Committee identified seven classes of experiments that it believes illustrate the types of endeavors or discoveries that will require review and discussion by informed members of the scientific and medical community before they are undertaken or, if carried out, before they are published in full detail. These categories represent experiments that are feasible with existing knowledge and technologies or with advances that the Committee could anticipate occurring in the near future. Some of them represent the types of naturally occurring genetic changes in pathogens that have led to disease pandemics such as the "*Spanish Flu*" in 1917-1918 or the recently recognized disease "*severe acute respiratory syndrome*" (SARS) but which could now be engineered in the laboratory. Others have been part of the history of biowarfare research and development. Furthermore, carrying out these types of experiments could, in some instances, lead to the potential for great damage without significantly advancing our knowledge in ways that would either greatly increase our ability to defend against them or our ability to promote human health by preventing, diagnosing, or treating common human diseases. The concerns deal with infectious agents or their products because the Committee believes that self-replicating agents or their products pose the most imminent biological threat.

The seven areas of concern listed here only address potential microbial threats. Of course, modern biological research is much broader, encompassing all of the health sciences, agriculture and veterinary science. It also includes diverse industries such as those that manufacture pharmaceuticals, cosmetics (e.g., Botox), and soft drinks (e.g., citric acid production). Moreover, all of these areas are changing rapidly. The great diversity as well as the pace of change makes it imprudent to project the potential both for good and ill too broadly and too far into the future. Therefore, the Committee has initially limited its concerns to cover those possibilities that represent a plausible danger and has tried to avoid improbable scenarios. Over time, however, the Committee believes that it will be necessary not only to expand the experiments of concern to cover a significantly wider range of potential threats to humans, animals or crops but also to include oversight of work conducted for or performed within the private sector as well as non-NIH government facilities and sponsored activities that are not already voluntarily complying with the Guidelines.

Experiments of concern would be those that:

1. *Would demonstrate how to render a vaccine ineffective*. This would apply to both human and animal vaccines.
2. *Would confer resistance to therapeutically useful antibiotics or antiviral agents*. This would apply to therapeutic agents that are used to control disease agents in humans, animals or crops. Introduction of ciprofloxacin resistance in *Bacillus anthracis* would fall in this class.
3. *Would enhance the virulence of a pathogen or render a nonpathogen virulent*. This would apply to plant, animal, and human pathogens. Introduction of cereolysin toxin gene into *Bacillus anthracis* would fall into this class.
4. *Would increase transmissibility of a pathogen*. This would include enhancing transmission within or between species. Altering vector competence to enhance disease transmission would fall into this class.
5. *Would alter the host range of a pathogen*. This would include making nonzoonotics into zoonotic agents. Altering the tropism of viruses would fit into this class.
6. *Would enable the evasion of diagnostic/detection modalities*. This could include microencapsulation to avoid antibody based detection and/or the alteration of gene sequences to avoid detection by established molecular methods.
7. *Would enable the weaponization of a biological agent or toxin*. This would include environmental stabilization of pathogens.

***Review process***

The NIH Guidelines require creation of an Institutional Biosafety Committee (IBC) when research is conducted at or sponsored by an entity receiving any NIH support for recombinant DNA research. Most of the 400 or so IBCs registered with NIH are at institutions that are subject to the NIH Guidelines and for whom IBC registration is mandatory. While most of these institutions are academic, some industry-based IBCs are registered with NIH as a consequence of receiving NIH support. In other instances, companies voluntarily comply with the NIH Guidelines as a means of demonstrating a commitment to a "*gold standard*" for safety practices. Several Federal agencies and laboratories have made compliance with the NIH Guidelines a condition of their support of intramural and extramural research projects. Furthermore, a number of federal IBCs are registered with NIH.

All the experiments that fall within the seven areas of concern should currently require review by an Institutional Biosafety Committee (IBC). This review should take place regardless of the source of funding, whether the institution doing the research is public or private, and whether it is a university, government laboratory, or business. We thus recommend relying on the system of IBCs as the first review tier for experiments of concern. We note that funding agencies also have a potentially important role to play in flagging experiments of concern at the proposal review stage.

Like the broader life sciences community, the members of the IBCs will require substantial education in the potential risks associated with advanced biotechnology research in order to handle this task competently. Many IBCs may need to add expertise in immunology, virology, pathology, and epidemiology to undertake this new responsibility. Some of this is already occurring as part of implementing the requirement of the Bioterrorism Response Act, but more will need to be done. To ensure the most consistent application possible of the review process—and as a reassurance to the research scientists subject to the new review—regular opportunities for members of IBCs to gather and discuss the process should be provided on a continuing basis.

We recommend that the form researchers now use to submit their experimental designs to the IBC be amended to include another category where researchers would designate whether, in their judgment, their proposed projects fit into an area of concern. The IBC would then review that issue along with the other aspects of the project that it is evaluating, carefully weighing potential benefits versus potential danger. Occasionally, the IBC may discover that what is proposed is forbidden under current guidelines, and would not approve the research. In most cases, however, it would either designate the project as acceptable to move forward or as one raising concerns that need further consideration.

Experiments that need further consideration would be referred to an expanded NIH RAC and possibly the Director of the National Institutes of Health for approval or denial of permission to proceed with the proposed experiment. We recommend this route because so many of the experiments in the area of concern would fall under the purview of the RAC already and because it has an established track record of facilitating research while protecting public safety. We propose that when the RAC takes up this new duty, it initially translate our categories of experiments of concern into a set of guidelines for IBCs to use. It should then improve and update these guidelines as needed as its experience with the process grows. The RAC will need substantial new resources to take on this additional task, and both it and the IBCs may need to incorporate new expertise to handle the task.

Under our recommendation, the RAC would begin to review some projects in the areas of concern from all relevant research institutions. This would be a substantial expansion from its current jurisdiction over research funded by the NIH and those institutions that comply voluntarily. The RAC guidelines would thus need to be revised and reproposed in the Federal Register to reflect this expanded scope and mandate. As we envision this review of the experiments of concern, when an IBC refers a project to the RAC, the RAC would carefully weigh the potential benefits and dangers of the project, and come to its own independent judgment. The RAC may approve some projects referred by IBCs to go forward at this point, recommend that the research not be undertaken, or that modifications be made to the research design to minimize the potential risks.

**Review at the Publication Stage**

We recommend relying on self-governance by scientists and scientific journals to review publications for their potential national security risks.

By the time a manuscript is submitted for publication, substantial information about the research may have already been disseminated through informal professional contacts, presentations of preliminary results at scientific meetings, or consultations with colleagues. This is why the committee recommends a system that can address research at its earliest stages, and why it is so important to make scientists aware of their personal responsibilities to consider the balance of risks and benefits in research they consider undertaking. Nevertheless, publication of research results provides the vehicle for the widest dissemination, including to those who would misuse them. It is thus appropriate to consider what sort of review procedures can be put in place at the stage of publication to provide another layer of protection. The Committee believes strongly that this part of the system should be based on the voluntary self-governance of the scientific community rather than on formal regulation by government.

Proposals to limit publications have caused great concern and controversy among both scientists and publishers. The norm of open communication is one of the most powerful in science. To limit the information available in the methods section of journal articles would violate the norm that all experimental results should be open to challenge by others. But not to do so is potentially to provide important information to biowarfare programs in other countries or to terrorist groups.

Journals in the life sciences have already responded to the challenge in a variety of ways; the procedures that a number of the leading publications have undertaken to screen manuscripts. The joint statement by editors of four major journals in the life sciences issued in February 2003 was a major step toward developing this part of the system of oversight the committee believes will be necessary. It was also an important example of the ability of the scientific community to address the potential risks of its activities.

Ultimately, any process to review publications for their potential national security risks would have to be acceptable to the wide variety of journals in the life sciences, both in the United States and internationally. The Committee believes that continued discussion among those involved in publishing journals—and between editors and the national security community—will be essential to creating a system that is considered responsive to the risks but also credible with the research community. The national advisory board recommended in the next section could serve as a forum for such discussions and for creating greater consensus in the scientific community about the appropriate role of and process for review at the publication stage.

On the broader question of classification, the committee believes that the principle set out by the Reagan Administration in 1985 in National Security Decision Directive 189 remains valid and should continue to be the basis for U.S. policy. The policy states that: "to the maximum extent possible, the products of fundamental research remain unrestricted. ... where the national security requires control, the mechanism for control of information generated during federally-funded fundamental research in science, technology and engineering at colleges, universities and laboratories is classification." The committee's support for self-governance by the scientific community through appropriate reviews by journals and other publication outlets should not be construed as endorsing the creation of "*sensitive but unclassified*" information in the life sciences. We believe that the risks of a chilling effect on biodefense research vital to U.S. national security as the result of inevitably general and vague categories is at present significantly greater than the risks posed by inadvertent publication of potentially dangerous results. A system of review based in scientific self-governance can, we believe, effectively address the security risks without discouraging scientists from participating in important biodefense research.

**Creation of a National Science Advisory Board for Biodefense**

We recommend that the Department of Health and Human Services create a National Science Advisory Board for Biodefense (NSABB) to provide advice, guidance, and leadership for the system of review and oversight we are proposing.

We recommend the formation of a National Science Advisory Board for Biodefense (NSABB) to provide advice, guidance, and leadership for the system of review and oversight we are proposing. The NSABB would serve a number of important functions for both the scientific community and the government. At the most general (strategic) level, it would serve as a point of continuing dialogue between the scientific community and the national security community and as a forum for addressing issues of interest or concern.

At the operational (tactical) level, it would provide case-specific advice on the oversight of research and the communication and dissemination of life sciences research information that is relevant for national security and biodefense purposes. Because of its important bridging functions, its members should include both leading scientists and national security experts, including those with experience in managing scientific research in federal agencies. Particularly in the early phases of its work, it would be desirable to include among the Board's members a few scientists or engineers from research fields long associated with applications to national security.

In terms of the regulatory aspects of the operation of our proposed system, we recommend that the Board periodically review and suggest updates to the "*Experiments of Concern.*" We also recommend that the Board review and suggest updates to the list of "*select agents*" and to policies regarding the international exchange of biological agents. A review of the select agents list by DHHS is already required every two years but the Board could serve a useful and important function by providing an independent assessment as an input to that process.

For the system's self-governing phases, we recommend that the NSABB serve as a resource. This could include aiding the professional societies in developing education programs, as well as providing a convening mechanism. It could also include assisting those producing publications in the life sciences. The Board could provide a convening mechanism for journal editors, organizing periodic discussions among them as they develop and evaluate their review processes. The Board could review and comment on proposed procedures on request, and perhaps serve as a clearinghouse so that journals that have not already adopted review procedures could have ready access to examples of what others are doing.

It would be very important for the Board to reach out beyond the United States to the many international publications in the life sciences and to find ways to include their leaders in discussions. The Board might also provide advice on request about particular manuscripts that raise concern, perhaps by organizing small groups of experts to assess the trade-offs between the scientific merits of the research, especially that with the potential to advance knowledge relevant to biodefense, and the risks of publishing information that might assist terrorists or proliferatant states.

So far, we have only discussed the functions of the Board that relate to the potential risks of research in advanced biotechnology. But we also recommend that the Board have the capacity to advise the government on how the life sciences can contribute to alleviating the risks of bioterrorism and biological weapons through new research in areas such as vaccine and antibiotic development, new detection devices and technologies, and preventative public health measures. This advisory function would serve as a continuous reminder that any system of review and oversight must operate in ways that do not put the United States—and the world—at risk of losing the great potential benefits of biotechnology. Having a Board that was informed and aware of the latest research developments, even including manuscripts not yet published, would provide the capacity for "*early warning*," alerting the government to the risks of new findings or techniques that should be met by focusing research resources on appropriate responses or countermeasures.

We considered a number of options for the organizational location for the NSABB. There are clear trade-offs between an independent board that offers its advice to government and one that is a formal advisory body to one or more federal agencies. The relationship between the life sciences community and the national security community is new and still tenuous, with significant potential for suspicion and misunderstanding on both sides. The topics the Board will address are both sensitive and controversial within the scientific community, and there is a risk that a formal federal role could raise concerns about the NSABB's capacity to offer genuinely independent advice. Whatever the home, the organization that houses the Board must have high credibility with the scientific community, since its engagement is essential to success. But it must also be able to command the cooperation and trust of the national security community and of the full range of U.S. research facilities, public and private, and publications. A formal attachment to the U.S. government would ensure access to the relevant high-level decision makers. The host organization should also have sufficient international standing to gain the necessary cooperation from the research communities of other countries. Another important consideration is the suitability of the organization for conducting some closed deliberations, while overall maintaining transparency and public trust in the process.

No solution meets all the criteria, but on balance we believe that the logical organizational location for the NSABB is within the Department of Health and Human Services providing advice to the secretary of that Department. DHHS already has a leading role in biotechnology research, particularly that related to the Experiments of Concern. Location within the DHHS would also connect the Board directly to the other parts of our proposed system, the RAC and the IBCs, while not limiting its capacity to work with other relevant agencies or private groups. In addition, this approach would fit within the division of labour created under the Bioterrorism Response Act, where HHS provides tactical advice—as the NSABB would do on specific issues and cases—and the Department of Homeland Security is charged with formulating overall strategy. We note that the Board will require significant financial resources to carry out its responsibilities, although the Committee did not attempt to estimate an amount.

It would be important for the Board to monitor the development and operation of the system we recommend and perhaps of other processes that the government or private organizations may put in place as well. The substantial expansion of funding for research in biodefense now in progress and anticipated suggests that it will be vital to assess how these new resources affect the conduct of research and to be ready to make timely adjustments. The monitoring should be done with the goal of suggesting ways to improve the system's operation and efficiency. But it should also include the possibility of proposing that parts of the system be overhauled or even eliminated if they prove ineffective or an impediment to important scientific research.

As discussed further, international coordination and cooperation will be necessary to make any effort to mitigate the risks of bioterrorism effective. Therefore, the Committee believes that the establishment of an NSABB within the United States can serve as the basis for international dialogue aimed at reducing the risks of subversion of legitimate life sciences research efforts. Review systems, comparable to the one proposed involving the IBC and RAC, already exist in many nations. These were established as an outgrowth of the Asilomar conference in 1975. In the same manner, other countries should be encouraged to establish counterparts to the NSABB so that the community of life scientists globally can work together to reduce the risks of the offensive applications of life sciences research.

**Additional Elements for Protection Against Misuse**

We recommend that the federal government rely on the implementation of current legislation and regulation, with periodic review by the NSABB, to provide protection of biological materials and supervision of personnel working with these materials.

The major focus of the Committee's work has been reviewing the adequacy of the current U.S. regulatory system to deal with the increased concerns about misuse of research in advanced biotechnology and recommending a system that could better address those risks. But there are other elements of the current regulatory system that the Committee believed should be reviewed and evaluated because of their important impact on the conduct of research.

***Physical containment***

Absolute containment of organisms with potential for bioterrorism is not a realistic expectation. Many of these agents can be cultured directly from nature or obtained from small animals available at any pet store or exotic animal "*swap meet*;" no genetic modification is required to convert them into weapons. It is, therefore, not feasible—with the possible exception of smallpox—to prevent knowledgeable individuals from obtaining any of the agents listed on the CDC select agent list by simply increasing the physical security of the laboratory environment.

There may, however, be individuals or rogue groups who lack the expertise either to isolate or grow pathogenic organisms, suggesting that cost-effective efforts should be made to limit access to them. Safeguarding the collections of existing agents is an obvious priority that in large measure has been addressed through recently passed legislation and implementing regulations. The CDC's and APHIS's designation of certain pathogens as "*select agents*" is an appropriate starting point for identifying strains and isolates that need to be secured. Additional agents, some of which have only recently been isolated, could be added to the list; agents might also be removed from the list if their potential for misuse is no longer considered a serious risk. Appropriate regulations should be enforced through the existing institutional biosafety committees.

It is crucial to avoid well-meaning but counterproductive regulations on pathogens. For example, regulations that force or provide a strong incentive for scientists to purge archival stocks of human, plant, and animal pathogens may deprive of us material that could be critical in the forensic identification of intentionally introduced pathogens into our environment and mounting an effective defense against bioweapons agents. A similar caution exists in assessing the risks of handling DNA fragments from select agents, which may pose no potential risk at all. Rules for containment and registration of potentially dangerous materials must be based on scientific risk assessment and informed by a realistic appraisal of their scientific implications. Moreover, scientific input is essential to ensure that these rules are clear as well as responsive to periodic assessment of the current technologies and capacities.

We recommended above that the NSABB review the security designations of biological agents. The Board could also be available to provide advice on short notice about revising regulations in response to new developments. Rules governing transfer of materials between laboratories to prevent unauthorized distribution or diversion might also be regularly reviewed by the NSABB so that new threats could be recognized and responded to and unnecessary impediments identified for removal.

***Trained personnel***

In some areas of technology, the limiting ingredient is the existence of trained personnel. There are two aspects to the technical expertise: general microbiological know how and knowledge about how to weaponize bacteria or viruses. General microbiological training sufficient for culturing and growing pathogenic microorganisms at levels of significant concern is available in high school and first year college biology courses; majors in microbiology would be sophisticated enough to grow many select organisms. It should be remembered that the procedures used to grow pathogenic bacteria are identical to those used for harmless bacteria, differing primarily in the need for precautions to ensure the safety of the workers. Moreover, training in basic microbiology is widely available outside the United States. Efforts to identify or control knowledgeable personnel within the United States are, therefore, impractical, and surveillance of such personnel would not, in our opinion, offer much security.

The procedures for admitting foreign students and scientists to the United States for study and collaborative research must reflect the importance of keeping universities as open educational environments. This must be weighed against national security concerns for limiting the spread of information to adversarial groups and admitting individuals who pose risks to domestic security. Establishing procedures for limiting visas is the role of the Department of Homeland Security and allied government agencies; universities and research scientists may provide information, but they have no investigative powers. It should be borne in mind that scientists and students, in particular those from developing countries, are likely to have a major interest in infectious diseases, because such diseases impose a devastating health burden in their home countries. Also, having colleagues and well-trained health workers in other countries increases U.S. security by enabling early access to information about emerging infections.

In June 2003 the presidents of The National Academies responded to the growing concerns that new security measures directed against foreign students, workers, and scholars could cause potentially serious damage to the conduct of science in the United States by issuing a statement that read in part:

To make our nation safer, it is extremely important that our visa policy not only keep out foreigners who intend to do us harm, but also facilitate the acceptance of those who bring us considerable benefit. The professional visits of foreign scientists and engineers and the training of highly qualified foreign students are important for maintaining the vitality and quality of the U.S. research enterprise. This research, in turn, underlies national security and the health and welfare of both our economy and society. But recent efforts by our government to constrain the flow of international visitors in the name of national security are having serious unintended consequences for American science, engineering, and medicine. The evidence we have collected from the U.S. scientific community reveals that ongoing research collaborations have been hampered; that outstanding young scientists, engineers, and health researchers have been prevented from or delayed in entering this country; that important international conferences have been canceled or negatively impacted; and that such conferences will be moved out of the United States in the future if the situation is not corrected. Prompt action is needed.

**A Role for the Life Sciences in Efforts to Prevent Bioterrorism and Biowarfare**

We recommend that the national security and law enforcement communities develop new channels of sustained communication with the life sciences community about how to mitigate the risks of bioterrorism.

By signing and ratifying the Biological and Toxin Weapons Convention, the United States renounced the use and possession of such offensive weapons and methods to disseminate and deliver them. Given the increased investments in biodefense research in the United States, it is imperative that the United States conduct its legitimate defensive activities in an open and transparent manner. This should clear the way for all biomedical scientists to contribute to the development of defensive measures that would mitigate the impact of the use of such weapons against people, plants, and animals. For the scientific community to be a willing partner in biodefense research, there must be trust and understanding between the scientific community and the defense, intelligence, and law enforcement branches of government.

The recent experience with anthrax dispersal in the United States made clear that there are individuals or groups in the world who will use the most horrific weapons, including pathogenic organisms, to kill innocent people for vague and unstated political goals. Added to the already existing concern about nonstate actors seeking BW capabilities, this has put bioterrorism along with biological warfare on the front burner for both the military and civilian populations. It has also meant that groups of people who had little history of working together, such as basic biomedical scientists and the FBI and CIA, must now find a way of sharing information and expertise. The nuclear physics/ Department of Defense community, which grew from a relatively small group during World War II, has had a long history of participation with intelligence and defense. Biomedical science, as already discussed, has had a different history. The intelligence and law enforcement agencies need the academic scientists both for the expertise they might provide about the nature of current agents and the potential for new ones and for the best advice on limiting the spread of new technologies that would make countermeasures even more difficult. It might be desirable for components of the national security and law enforcement communities to establish advisory boards of basic scientists and clinicians with expertise in specializations such as viral disease, bacterial pathogens, biotechnology, immunology, toxins, and public health, as well as others in the area of basic molecular biology. These advisory boards could help members of these communities keep current in relevant areas of science and technology and provide trusted set of advisors to answer technical questions.

**Harmonized International Oversight**

We recommend that the international policymaking and scientific communities create an International Forum on Biosecurity to develop and promote harmonized national, regional, and international measures that will provide a counterpart to the system we recommend for the United States.

Any serious attempt to reduce the risks associated with biotechnology must ultimately be international in scope, because the technologies that could be misused are available and being developed throughout the globe. A number of countries and regional and international organizations are already moving forward to develop programs and policies on aspects of the problem; the initiatives include consultations among the parties to the BWC on best practices for the security and oversight of pathogens and toxins. These approaches must be harmonized and widely adopted in order for them to be effective. Just as the scientific community in the United States must become deeply and directly engaged, the commitment of the international scientific community to these issues is needed to implement the recommendations contained in this report. Diverse groups of scientists, academicians, and policymakers must be brought together for a sustained dialogue in order to develop consensus and devise a path forward.

We do not expect our recommendations to provide a "*roadmap*" that could simply be adopted internationally without significant modifications or adaptations to local or regional conditions. But any effective system should include all the issues addressed by our recommendations. The committee therefore recommends, as a next step, convening an "International Forum on Biological Security" to begin a dialogue within and between the life science and the policymaking communities internationally. Among the topics for this international forum are:

1. Education of the scientific community globally, including curricula, professional symposia, and training programs to raise awareness of potential threats and modalities for reducing risks as well as to highlight ethical issues associated with the conduct of biological science.
2. Design of mechanisms for international jurisdiction that would foster cooperation in identifying and apprehending individuals who commit acts of bioterrorism.
3. Development of an internationally harmonized regime for control of pathogens within and between laboratories and facilities.
4. Development of systems of review to provide oversight of research, including defining an international norm for identifying and managing "*experiments of concern*."
5. Development of an international norm for the dissemination of "*sensitive*" information in the life sciences.

The committee believes that, to be most effective, this and other forums should be sponsored by international organizations with the standing and credibility within both the policymaking and scientific community. Different topics within this broad agenda may be more appropriate for different organizations. Potential sponsors could include the World Health Organization and the United Nations Educational, Scientific and Cultural Organization (UNESCO) as formal international governmental organizations with direct links to government policymakers. Among nongovernmental scientific organizations are the International Council for Science (ICSU), created in 1931 "to identify and address major issues of importance to science and society, by mobilizing the resources and knowledge of the international scientific community;" and more recently created organizations of the world's academies of science such as the InterAcademy Panel on International Issues (IAP) and the InterAcademy Council (IAC) that seek to bring the prestige and convening capacity of these bodies to bear on crucial international problems.

Finally, the committee notes the uncertain international foundation for authoritatively addressing the issues that we have considered. No international agreement addresses the potential threats posed by the misuse of research in the biological sciences, and no intergovernmental organization has relevant oversight authority to promulgate guidelines or procedures. The committee believes that convening an international forum to address these gaps demands international and interdisciplinary mobilization of resources and capabilities.

This report reflects the increasing attention being paid by scientists and policymakers to the potential for misuse of biotechnology by hostile individuals or nations and to the policy proposals that could be applied to minimize or mitigate those threats. The term "*misuse of biotechnology*" is a phrase that captures a wide spectrum of potentially dangerous activities from spreading common pathogens (*e.g.*, spraying *Salmonella* on salad bars) to sci-fi plots of transforming pathogens into the next "*Andromeda strain.*" Our Committee addressed one important part of this spectrum of risks: the capacity for advanced biological research activities to cause disruption or harm, potentially on a catastrophic scale. Broadly stated, that capacity consists of two elements: (1) the risk that dangerous agents that are the subject of research will be stolen or diverted for malevolent purposes; and (2) the risk that the research results, knowledge, or techniques could facilitate the creation of "*novel*" pathogens with unique properties or create entirely new classes of threat agents.

Throughout the Committee's deliberations there was a concern that policies to counter biological threats should not be so broad as to impinge upon the ability of the life sciences community to continue its role of contributing to the betterment of life and improving defenses against biological threats. Caution must be exercised in adopting policy measures to respond to this threat so that the intended ends will be achieved without creating "*unintended consequences.*" On the other hand, the potential threat from the misuse of current and future biological research is a challenge to which policymakers and the scientific community must respond. The system proposed in this report is intended as a first step in what will be a long and continuously evolving process to maintain an optimal balance of risks and rewards. The committee believes that building upon processes that are already known and trusted and relying on the capacity of life scientists to develop appropriate mechanisms for self-governance, while greatly expanding the consultation and dialogue between the science and national security communities, offers the greatest potential to find the right balance. This system may provide a model for the development of policies in other countries. Only a system of international guidelines and review will ultimately minimize the potential for the misuse of biotechnology.

## CONCLUSION

Professor of history Martin Van Creed has written that the rules of war exist to protect the armed forces themselves by defining a common culture code that differentiates an army from a mob. More importantly, he states that these rules carefully define how armies can and cannot be allowed to kill other combatants and noncombatants. These are in effect a societal rule that draws the line between murder and war. Countries that ignore the rules usually provoke punishment, often through annihilation or later war crime trials. These rules of war once covered the use of crossbows, submarine warfare, strategic bombing and mine laying; today of course, they still cover CB warfare.

Over the last century, nations have attempted to restrain the use of CB weapons under these special rules, though more from fear of retaliation than from moral distaste. Because of past association with strategic bombing theories developed between the world wars and after World War II, the advent of nuclear weapons, chemical and biological weapons have been put on the same scale as nuclear devices. Contrary to this view, CB warfare has never resulted in thousands or millions of innocent casualties, the nightmare that has derived the international arms control community to such treaties as the Biological Weapons Convention and chemical Weapons Convention. History has shown that nations continue to develop CB weapons because there are tactical advantages to be gained: when one combines CB weapons use with conventional military tactics against an enemy without a good defense, it dramatically reduces the time needed for victory. IN these days of billion-dollar conflicts, that gets attention. Increasing global proliferation of these weapons drives the need for defensive measures despite these special "rules" of war.

The chemical Corps survived to become part of the Army's twenty-first century military force, not by planned evolution but by world events that constantly demanded that the US military retain an ability to protect itself against CB warfare. In 1972 the Army turned its back on the Chemical Corps because the service's leaders did not see the operational benefits of retaining such combat support. As a result, a decade slipped by without modernizing CB defense equipment; without leaders understanding that CB warfare is a tactical, not strategic consideration; without CB defense specialists staffing combat units and command headquarters; and without the specialized chemical defense companies and biological detection teams supporting combat arms units. While the 1980s featured a great deal of improvement in doctrine development, leadership, education, chemical defense unit build-up and equipment modernization, the GAO report released in May 1991 pointed out the serious deficiencies in the Army prior to Desert Shield. Imagine, if you will the situation in August 1990 if General Creighton Abrams in 1972 had been successful in relegating the Chemical Corps to a special weapons function under the Ordnance Corps. More importantly, will the four services, having looked into the eyes of the dragon, now forget that the threat of CB warfare ever existed during the Gulf War? Or does the US military leader leadership now acknowledge the need for continued reform and further efforts? While the Chemical Corps obviously has room for improvement, its efforts are wasted until the three and four star Army leaders identify these challenges and aggressively support the Chemical Corps in meeting them.

**National Policy on NBC Warfare Requires Updating**

The US government's policy on NBC warfare has been one of "*no first use*" only since World War II, as the US senate had refused to ratify the Geneva Convention in 1925. Because of the potential threat from Germany and Japan, the US military scrambled for eighteen months to develop a retaliatory capability in the form of mustard and phosgene agent-filed munitions. President Franklin D. Roosevelt publicly stated the government's policy that under no circumstances would the United States resort to using CB weapons unless they were first used by the Axis. If used by Germany, Italy or Japan, the US military would retaliate against munitions centers, seaports and other military objectives.

After World War II, President Roosevelt's policy remained the unwritten rule for most administrations, although the US government had still not ratified the Geneva Convention. President Richard Nixon publicly reaffirmed the policy of no first use of chemical weapons and abandoned the use of biological warfare agents altogether. The Senate's final ratification of the Geneva Convention in 1975 did not change the US policy of retaliation in kind, using the current stockpile of chemical weapons. The re-initiation of the binary chemical weapons program in 1985, meant to replace the aging unitary chemical munitions, forced the Soviets to return to the negotiations table on the subject. In the 1990 Bush-Gorbachev talks, the US government abandoned the future production and storage of chemical weapons, receiving promises from the Soviet Union that it would do likewise.

Now as of April 1997, the broader, multilateral Chemical Weapons Convention treaty has come into play. The CWC treaty outlines a verifiable ban on all production, storage and use of chemical weapons and carries over 180 nations' signatories. This represents a great change from the past policy, of maintaining both a strong defensive posture and a viable retaliatory capability, to one of maintaining only defensive measures. The one policy we knew, which worked for over forty-five years—threatening retaliation by chemical munitions if our forces were attacked with *chemical weapons* - is gone. On the other hand, the CWC treaty affects all countries by, at the least, causing them to go to some effort to shield the size and potential effects of their offensive CB programs. The future threat for US forces will not be continuous hail of chemical rains. Rather, it will be infrequent and inconsistent use of chemical agents to delay and disrupt combat operations, intended to give the opposing force an advantage at key points in the battle.

The US military is at a dangerous transition. It has given up its retaliatory capability and defensive capabilities remain weak, while the CWC treaty's verification measures have not been tested. If our force are attacked with CB weapons, the military leadership will have to choose nuclear or massive conventional retaliation. In a recent war game sponsored by the DoD counter-proliferation office, the players explored a combined CBW scenario in the Persian Gulf: after US carriers were attacked by aircraft armed with biological agent spray tanks, and terrorists attacked Dhahran with anthrax bombs, the United States responded by using nuclear weapons against Baghdad. While this ended the war game, it did not represent a satisfactory or realistic conclusion. This game move did not acknowledge that a US president would have to approve the nuclear attack of a country's capitol, let alone address the implications of attacking an Arab attack against US forces, would the president be prepared for the consequences of a strategic nuclear strike against civilian and industrial targets? Ordering a smaller tactical nuclear strike against military targets is not an option these days; the DoD yielded up their tactical nuclear weapons years ago.

The more realistic option is the massive conventional response, which is also politically and militarily risky. Experience during *World War II*, Korea, Vietnam (and, according to many, the Gulf War) shows that massive conventional bombing produces very little political or military payoff for the effort, while increasing pilot and aircraft exposure to increasingly sophisticated air defenses. In any case, to minimize the threat from CB weapons (e.g., strike production facilities, storage sites, delivery systems), you have to find them first. The intelligence job of finding and identifying these targets has gotten more difficult, not less. An increasingly diverse global economy allows countries to order just about anything they want through dummy companies, academic institutions, and public organization. Taking a page from the superpowers, many countries are investing in deep underground shelters to protect their storage sites and their command, control and communication functions. While the United States can identify most of these facilities, not much short of a nuclear device can get at them.

Military forces cannot predict with 100 percent clairvoyance when or where the CB agent munitions will be used, even with extensive intelligence assets. This meant that US forces would need to retain a strict defensive posture against CB agent use at all times in a theater, increasing logistics demands and decreasing morale. Meanwhile, the enemy needs only a minimal investment in defensive equipment since it controls the time and location of chemical weapons release, thus minimizing effects on its forces. This lopsided arrangement actually increases the possibility that an enemy force would use CB agent munitions, since its chemical weapons will eliminate the US military's high-technology edge. Army studies prove that chemical warfare results in a drop in overall preparedness and effectiveness, anywhere from 25 to 100 percent, depending on the scenario and the troop's level of training. Adversaries using CW munitions will multiply their combat power by degrading the performance of opposing troops that lack defensive equipment or training. Given that current force structure decisions are putting smaller US forces against larger opponents, losing the high-technology advantage will result in being overwhelmed on the battlefield. If we are trying to minimize US casualties, this should not be an acceptable course of action.

Can the US military survive and sustain its forces on the future NBC battlefield through strictly defensive methods? In theory, yes—but it would call for careful planning, expensive stockpiling and extensive training. It requires an aggressive development program that provides state-of-the art defensive equipment. It requires continued reliance on chemical defense units that can decontaminate military equipment and large areas of terrain (such as ports and airfields), while keeping pace with mobile forces. That means maintaining active and reserve units that are ready, well trained and have worked with combat units, integrating CB defense into combat operations. It calls for an increased focus on how combat arms units can minimize the impact of persistent CB agents. By integrating chemical defense training with their mission-oriented combat training, soldiers could learn to fight on contaminated battlefield, making CB warfare irrelevant. This would enable them to retain most of their intended lethality, which today can be seriously degraded, as seen in the CANE tests.

Last and perhaps as important, we need to reexamine our investment in human intelligence sources. The US government needs to remain alert to new and changing dangers such as terrorist use of CB agents and emerging Third World threats. Yet our government has made little attempt to shift its vision in this respect from the *Cold War* to the "*New World Order.*" There are less than fifty CB intelligence analysts in the entire federal government, as compared to nearly ten times that number dedicated to nuclear weapons intelligence. As seen with the intensive research conducted for the Khamisiyah incident, there are few analysts that understand and can validate CB weapons effects. More are needed to track and verity offensive CB weapons programs in countries like Libya, North Korea, Iran and Iraq. Unless this imbalance is corrected, US forces will again have no idea what their opponents have in the way of CB weapons, where the weapons are located, or how to attack them.

The US military must reexamine its policy concerning enemy use of CB weapons, and find a solution that is practical and feasible. Because we are not facing a massive use of CB agents, we can no longer justify a righteous massive retaliation. If massive force is no longer a credible deterrent, US forces in the future will be as fear-stricken as they were when the Scuds filled the skies in January 1991—unless they have full faith in their defensive capabilities.

**DoD NBC Defense RDA Program is Still Recovering**

Several material areas sorely required attention during the Gulf War. The most evident was the need for improved airlift and sealift capabilities. While C-17 cargo planes, heavy trucks and merchant marine ships will never be as sexy or alluring as F-22 fighter aircraft, Abrams tanks and Seawolf submarines, there certainly was a strong case made when the military had to rely on commercial aircraft and ships older than the generals and admirals in order to get to the theater of operations. Of course, NBC defense equipment was another area of concern. Despite the Gulf War (or because of the lack of CB warfare during this conflict), CB defense equipment continues to struggle on as a small, relatively insignificant portion of the Pentagon's acquisition funds (less than one-fifth of 1 percent).

NBC defense equipment has always been criticized by both the military troops that use the equipment and the chemical specialists that train with it. Protective suits are too hot; protective masks are too restrictive; collective protection shelters are not adequate; decontamination apparatuses use too much water; CB agent detectors false-alarm too often; and medical antidotes can not be trusted. All these complaints came too late for the troops in Operation Desert Storm. These concerns were primarily a result of the combat arms sitting by the sidelines, allowing the Chemical Corps to determine what the combat troops were going to use. The Chemical Corps did deliver detectors that detected chemical agents, protective ensembles that would have saved lives, and decontaminants did work very well. The *BUT* part is that the combat arms units did not understand that getting CB defense equipment that works in only half the equation; there remains such matters as false alarms due to overly-sensitive detectors, hot burdensome suits, corrosive decontaminants. If the combat arms incorporated CB warfare considerations into their war gaming, they might see the benefits and challenges of CB defense equipment, and they might get more involved. This active involvement will become more and more important as the military gives up its chemical munitions and relies solely on defensive equipment.

As a result of Public Law 103-160, there is an increased emphasis on better management of the NBC defense program. With the evolution of the one-star command CRDEC into the two-star command US Army Chemical and Biological Defense Command (CBDCOM) in October 1993, the Army is attempting to put much of its NBC defense program authority under the major subordinate command. The commanding general of CBDCOM also "*wears the hat*" of Deputy Chief of Staff for Chemical Matters, Army Material Command. CBDCOM is charged to mange all NBC defense programs for the Army and to work with the other services to mange better the DoD NBC defense program. CBDCOM will manage chemical stockpile sties, to include Rocky Mountain Arsenal and several programs at Dugway Proving Ground, Pine Bluff Arsenal and other locations. This is not suggest that the Army has gone full circle, back to the single consolidated NBC defense program it had prior to 1962; CBDCOM still has no direct control over medical NBC defense programs or the chemical demilitarization program. But it does have a great deal of authority to request funds conduct studies and procure NBC defense equipment.

Public Law 103-106 also meant reorganization for the entire DoD NBC defense community. The four services have formed a Joint Service Integration Group (JSIG) for joint doctrine and training issues and a Joint Service Material Group (JSMG) for joint material issues, both chaired by the Army, and both including all services as equal partners. The JSIG resides at For McClellan, with the Chemical School's commandant as its chair. Its sister organization, the JSMG, is chaired by the commanding general of CBDCOM At Aberdeen Proving Ground. These groups officially started their functions in January 1995, both overseen by a DoD Joint NBC Defense (Atomic Energy) (Chemical Biological Matters) (or ATSD(AE)(CBM). The joint NBC Defense Board is a high-level, general-officers panel (including civilian acquisition members and officers and to sort our inter services conflicts. The new joint groups may be more effective than their predecessors, primarily since they make recommendations on the annual NBC defense budgets through a joint consolidation Program Operations Memorandum strategy. As the Golden Rule states, those who have the gold make the rules.

The other services still do not trust the Army to control the DoD NBC defense program entirely. The four services still have difficulties agreeing on basic doctrine, or on roles and missions for the future. The other services continue to fight tooth and nail to retain exclusive control over their own destinies in the NBC defense program areas. Each service had its own offices, programs and "*rice bowls*," specifically for NBC defense ever since DoD gave them leeway to initiate their own NBC defense programs in 1960, programs that became further entrenched when the Chemical Corps was down-scaled in 1972. These two joint groups have the budget authority and congressional language to make the tough decisions necessary to reform the DoD NBC defense program, but this does not guarantee any quick or near-term improvements. Existing bureaucracies always fight change, and it will be some time before we see results.

One key to developing better NBC defense equipment may lie in using computer models and simulations to understand CB agent effects on military troops, units and equipment during combat. The Army's "*Battle Lab*" program at six TRADOC schools uses advanced models, simulations and war gaming tools to test new concepts of warfighting and to determine future program requirements. The Dismounted Battle Space Battle Lab at Fort Benning initially had the responsibility for "*weapons of mass destruction*," with NRDEC as its partner, but it focused almost exclusively on individual protection efforts. Other labs, such as the Mounted Battle Space Battle Lab at Fort Knox, the Depth and Simultaneous Attack Battle Lab at Fort Still, and the Early Entry Lethality and Survivability Battle Lab at Fort Monroe, have smaller, but not insignificant, roles for NBC warfare aspects. The Engineer

School and Center initiated its Maneuver Support Battle, Lab, which has taken over the champion role of working "*weapons of mass destruction*" issues, working with the Chemical Corps as it moves to Fort Leonard-Wood. It will be up to these Battle Labs to revisit the issue of NBC warfare on the modern battlefield. To date, there are no proven models of how CB agents affect troops and equipment, and therefore, there is no way to show senior leadership the benefits (or detriments) of fighting with a certain type of CB defense equipment. This is vital to assess the "*value-added*" of these programs. If warfighters, NBC defense subject-matter experts, and industry cannot gain a mutual understanding of NBC warfare, there will be no acceptance or understanding of modern CB defense equipment.

**CB Agents and Domestic Terrorism**

Terrorist organizations have taken an interest in CB weapons since at least 1972. Prospects for CB terrorism have increased with the international spread of dual-use technologies, increased access to published sources on producing CB agents. US intelligence sources have noted Moslem terrorist organizations conducting chemical defense training at their camps. Simple biological agents such as botulinum toxin can be developed with a moderate amount of education and equipment. They are inexpensive to procure and culture, easy to hide, require far less material than chemical weapons and are very easy to employ in an indiscriminate target-rich area such as a city. Most biological organisms, however, are more difficult to develop, store and disperse safely than chemical agents, lending to a pattern of terrorist groups external to the United States using chemical agents rather than biological agents. This could change, given today's global accessibility of technical and scientific resources.

Chemical and biological agents are inexpensive, easily understood, easily controlled, and easily disperse under a single terrorist's control. As the world's nations have not condemned the use of chemical agent weapon use in Yemen, Afghanistan, Laos, Iraq, etc., terrorists may feel that they can safely use CB agents rather than resort to nuclear devices. Some past examples of CB terrorism include:

1. In 1974, Muharem Kerbegovic was arrested in Los Angles after mailing toxic materials to a Justice of the Supreme Court and threatening to kill the president with nerve gas;
2. In 1980, the Paris police raided a Baader-Meinhof safehouse and discovered a culture of clostridinium botulinum (used to make botulin toxin) in a home laboratory.

3. In 1984, the Rajneesh cult in Oregon was accused of using salmonella to contaminate salad bars in local restaurants in an attempt to influence local electrons;
4. In 1991, German authorities thwarted a neo-Nazi plot to pump hydrogen cyanide into a synagogue.
5. In 1992, the FBI arrested two members of the Patriots Council in Minnesota for possession of less than one gram of ricin, under the Biological Weapons Anti Terrorism Act of 1989; and
6. In 1996, a former Aryan Nation member was arrested for mail-ordering three vials of bubonic plague to his home by giving a false lab code number to the American Type Tissue Collection.

The use of sarin in Tokyo subway system on March 20, 1995, illustrated how easy it is to employ chemical agents in an open society, and how much media attention a small, previously barley noticed group can receive as a result. Between 7:00 and 8:30 in the morning of March 20, several individuals left at least four containers wrapped in newspaper, disguised as lunchboxes, on various trains of three subway lines. Some were placed on hat racks, others on or under the seats. These containers were punctured, allowing its contents to evaporate. Shortly afterward, people riding the three lines bean coughing and complaining of headaches, diminishing vision, and nausea, classical nerve agent poisoning signs. Some collapsed to the floor, and as the trains halted at the next stop passengers rushed out, screaming for help and collapsing on the ground.

By 8:15, police were receiving emergency calls from sixteen subway stations. Tokyo's Metropolitan Police Department dispatched 11,000 police and rescue workers to assist the affected passengers. One assistant station manager, Kazumasa Takahashi, went abroad one train to remove a plastic bag holding a leaking container. He died soon afterward, along with seven others. In all, twelve individuals died, another fifty-four were in critical condition, and more than 5,000 were treated for related symptoms. By 8:45, service on all three subway lines was stopped, with twenty-six stations closed. About three hours after the first attack, the authorities realized they were dealing with a nerve agent. They moved into full response mode, sending police, firefighters and medical response teams into the subways.

The police targeted the Aum Shinrikyo, or Supreme Truth, a Buddhist sect that recruited dissatisfied college-educated citizens, as the probable perpetrators. Several of their members were arrested in Japan, and the police seized bulk precursor chemicals, including phosphorus trichloride, sodium fluoride and isopropyl alcohol. In the sect's building police discovered behind false walls industrial equipment for preparing bulk chemical mixes. As their leader, Shoko Asahara, explained a few days later, these are chemical that can be used for commercial purposes. They also are the prime ingredients for sarin; in fact, the Chemical Weapons Convention specifically names phosphorus trichloride as one of the dual-use chemicals on its "schedule 3" list.

The panic and concern over the next months was largely due to the fact that the Japanese police had been completely unprepared to react to this new form of terrorism. Perhaps more frightening was failure of intelligence agencies or police to predict this attack. There were a number of related cases that stretched back nearly two years, beginning with a case on July 2, 1993, when over 100 residents in Tokyo's Koto district complained of noxious white fumes rising from building owned by the sect. One year later, on June 21, 1994, seven people died and more than 200 were sickened by sarin fumes in the nearby town of Matsumoto. It was later discovered that Aum Shrinkiyo cultists had used a truck-mounted agricultural generator to target three judges in an apartment building (they were ruling on a local real estate issue involving cult members). On September 1, more than 230 people in seven town in the state of Nara (in western Japan) suffered from rashes and eye irritation from unknown fumes. On December 1994, material by-products of sarin were discovered in Kamikuishiki; January 4, 1995—the Aum Shrinkiyo filled a complaint accusing a company president of spreading sarin into its facilities in Kamikuishiki; March 5—eleven people were taken to a hospital after nineteen individuals inhaled noxious fumes in a train car in Yokohama (a test case?). On March 15, police discovered three attache cases containing an unknown (but nontoxic) gas. Each case held small motorized fans, a vent and a battery. Despite all this evidence the March 20 episode came as a surprise to the Tokyo emergency responders.

Experts still wonder why more subway riders did not die, considering the high lethality of the nerve agent. One theory rests on these possible speculations. First, the sarin mix was not pure. Later analysis showed that the chemicals were only 30 percent sarin by composition and were therefore not as lethal as they could have been. This design may have been deliberate, contrived to permit the couriers time to flee. Second, the delivery system was crude. Accounts told of the suspected couriers rupturing the one-quart containers immediately before departing the trains, allowing natural seepage to initiate the f... Not the most efficient process; a portable fan next to the p...e would have spread more quickly the fumes throughout the car... hird sarin is what is known as a *semipersistent* chemical agent—it h... about the same consistency as water, taking up to hours to evaporate completely. This would explain why the only individuals who died were the ones traveling in the same cars as the containers like the assistant station manager who picked up the package to remove it from the train. Last, the trains may have had a positive pressure ventilation, using fans to circulate air out of the trains and into the subway tubes. This precaution is a normal engineering method to keep gasses or smells in the subway tubes from intruding into the individual cars; the District of Columbia Metro, for instance, has this feature. In all the Japanese citizens were very fortunate the most survived the physical effects of this deadly agent.

Another subway in Yokohama was attacked a few weeks later on April 19, sending nearly 600 Japanese sent to the hospital for treatment of what was suspected to be exposure to phosgene gas. While this agent is not immediately suspected to be exposure to phosgene gas. While this agent is not immediately lethal, it can kill as the gas hydrolyses within the victim's lungs over a period of one to two days. Many people do not realize that this World War I chemical warfare agent is also a legitimate industrial chemical, used in the textiles and clothing industry as well as plastics manufacturing. As such bulk quantities of the chemical travel across the country every day in tankers on rail cars and trucks; there is no need for illicit manufacture phosgene, it is a very simple chemical compound to synthesize in the laboratory. A third suspected gas attack at a grocery store two days later caused twenty-five people to seek treatment. Because the attack occurred outdoors, the lethal chemical agents may not have their full potency. On May 6, the police defused a binary device that would have again released hydrogen cyanide into a Tokyo subway station.

The Aum Shrinkiyo may have been coming after American citizens as well. A task force composed of representatives of the FBI, Los Angeles police, the Federal Emergency Management Agency (FEMA), Environmental Protection Agency, public Health Service, Centers for Disease Control, and CBDCOM's Technical Escort Unit traveled to Los Angeles airport on April 13 to intercept two Japanese men associated with the cult. These men had in their possession written instructions on how to make sarin and a videotape that indicated an intended attack against Disneyland. Authorities at Disneyland had received a tip that a possible terrorist incident was going to take place over Easter weekend, when their attendance was expected to be higher than average. Although details are still vague, it appears that the terrorist group was planning to release sarin within one of the many theaters of exhibition halls at Disneyland. Because of quick reactions based on the tip, these men were apprehended, and America's version of Tokyo's experience was averted—for now.

Since March 1995, federal agencies have banded together under a program titled "*Domestic Preparedness*," to prevent and respond to acts of CB terrorism conducted within the United States. The FBI leads what is termed "*crisis management*" (what to do prior to a CB terrorist incident), while FEMA leads "*consequence management*" (what happens after the event occurs). Forces were in place at the Atlanta Olympics and the 1996 national political conventions, with a mission to prepare for any sign of CB terrorism. While it was an interim agreement among federal government agencies to cooperate prior to 1996, the Domestic Preparedness program has since been codified by executive order and public law and is under way. This program had just been started when it had its first acid test. Preplanning for large events can be accomplished, but the following case showed the real dangers.

In Washington, DC, on the evening of April 24, 1997, someone left a smelly manila envelope in the mail room of the B'nai Brith offices. Two employees opened the envelope; it contained a petri dish filled with an unknown, red, jelly-like substance and markings that suggested it held anthrax. The police took immediate steps to cordon off square blocks around DuPont Circle, an area that included the Australian Embassy and several traffic commuter routes. Firefighters were called in, as well as the FBI and CBDCOM's Technical Escort Unit. They evacuated the dish, an inside the building, 108 people were quarantined for nearly eight hours as medical technicians at the Bethesda Naval Medical Center analyzed the compound. Meanwhile, the two employee declared they were feeling sick. They along with twelve firefighters' were decontaminated with a chlorine solution.

As things turned out, there was no anthrax in the petri dish. While FBI and city officials praised the firefighters' quick reaction, it was a major overreaction to a minor threat. First, airborne anthrax is the typical biological warfare agent used, due to its much higher lethality than percutaneous anthrax. Since the sample was in a liquid-gelatinous substance, there was no chance of its escaping into the atmosphere. A two hundred-meter safety radius would have been as good as several square blocks; had this actually been anthrax there would have been very little threat to the surrounding area outside the building itself. Second, any biological agent needs time to incubate in a human host prior to causing sings of sickness. Certainly any symptoms felt by the two employees and the firefighters had to be psychosomatic. Exposed personnel could have been treated with vaccines prior to any real dangers to their health. The point here is that, left to their own resources, emergency responders will over-react to CB agents and make the wrong decisions unless they learn more about this area.

The Washington case was a free trial-run for federal counter-terrorism forces and city emergency response forces. It pointed out several challenges: the need for CB defense equipment responders, the need for an expert source of information around the clock and the need to train and educate local emergency responders in how to identify and manage the incident. Trained federal forces will be hard pressed to respond to unplanned emergency events in less than eight to twelve hours. This puts the burden on police, firefighters, emergency medial services, and 911 operators to know how to deal with the threat. It may be that this specialized training and equipment is the best local responders can do is to keep from becoming casualties themselves. If we expect that all the major metropolitan areas require this kind of expertise, a major investment of funds, equipment and time will be required, of which time is probably the most critical.

While the FBI and FEMA have traditionally led counter terrorism and disaster preparedness missions, the expertise on CB agents is in the Department of Defense. This has led to the development of an interagency federal task force and several programs for federal support to metropolitan emergency responders. An ambitious training program has targeted 120 major cities in the US, training their emergency responders to identify and respond correctly to CB agent terrorism. It will take some years, but this interagency task force envisions a certain degree of expertise at the local and state levels that will at least ensure that firefighters, police and emergency medical technicians are not killed rushing into a CB terrorist incident. But it should be understood by all that this will not guarantee against a similar Tokyo subway incident in the US—all it does is minimize the ensuing chaos and the casualties among responding professionals. As in any terrorist bombing incident, few things can minimize the number of civilian casualties other than interdicting the terrorist group itself prior to the event.

There are a number of additional cases, some still classified, where terrorist use of CB agents in the CB agents in the United States has been only narrowly averted. One reason why terrorists have not adopted these weapons more readily in these already unstable times may be the argument of cost versus availability. The 'bang for the buck" comparisons shows that CB agent kill their victims much more cheaply per casualty than bullets or explosive do. Bullets and explosives, on the other hand, are far more available on the open market than CB agent, and they are still effective attention-getters. As long as a terrorist group can achieve its goals with conventional weapons, it will probably do so.

These days, terrorist incidents compete with drive-by shooting, mass suicides, uprising, and international incidents for front-page news. One day soon, exploding a van under the World Trade center resulting in six deaths may not receive much attention. Worse yet, could the next federal building be attacked with CB agents instead of a fertilizer-based explosive? Or will terrorists attack key industrial sites to reproduce the next Bhopal, *Exxon Valdez*, or Chernobyl catastrophe? Is the United States ready to combat domestic terrorism in the form of CB agent attacks? AS one of the most open nations in the world, our society is very vulnerable. One fact remains: in the midst of news stories recounting the falling yen, a major earthquake, and North Korean demands for nuclear reactors, worldwide publicity catapulted the name Aum Shinrikyo, the Supreme Truth, to the homes of billions across the glove within one week. That alone should prompt concerns that others will seek to imitate its efforts.

**There is a Future for the Chemical Corps**

Future threats are not diminishing in number of volume, despite the successful completion and signing of the Chemical Weapons Convention in 1993. In every corner of the world, particularly in historical and regional conflict areas, proliferation of CB warfare programs continues. Many countries begin with production of industrial chemicals, move into first-generation (*World War I*) chemical agents, rapidly progress into development of second-generation chemical agents (*World War II* and later), and last, biological agent production. While the CWC will continue to put pressure on these countries, the extreme liquidity of the global economy, and the increasing sophistication of other nations' arms programs will make it increasingly difficult for the United States and its allies to control this proliferation. We can be assured that the next major conflict will include the use of CB agent munitions, in an attempt to resolve a conflict before its high price can take its toll and before major powers can be involved.

The US military's future is undergoing its own evolvement, as seen with the Quadrennial Defense Review (QDR) and its successors. With a diminished manpower and budget, a decreasing involvement overseas, and an increasing reluctance on the part of the American public to suffer any casualties in an overseas conflict, the military is searching for a new, defined role. The QDR's language concerning weapons of mass destruction, asymmetric, threats and the threat of ballistic missiles shows that DoD understands the danger exists. Military leaders just have not figured out exactly how they want to combat them. The lack of CB warfare during the Gulf War has caused more questions and theories than would have been the case if the Iraqi military had used these agents. Some emphasis focuses on a "21st Century Land Warrior" program combining protective clothing, mask and integrated CB agent detectors into the troops' combat ensembles. Others insist on more proactive measures, such as Theater High Altitude Air Defense systems and planes with high-powered lasers designed to defeat CB agent warheads on ballistic missiles. The main weakness is a continued search for a technological "fix" for a problem, a search that ignores the need for integrated NBC defense training and trained chemical specialists in future combat forces. The military still needs to test these concepts, using validated computer models that mimic the physical and physiological effects of NBC agents on troops and units, to understand the costs and benefits of these approaches.

Despite their close brush in 1991 with CB warfare, the four services did not immediately acknowledge the need to reform their roles and responsibilities in the DoD NBC defense program. Although the Gulf War division and corps commanders are in high positions throughout the Army today, their near experience with CB weapons use has not trickled down to the new generation of commanders. Despite the worry prior to the conflict over these weapons, some still feel that conventional firepower alone is more than adequate to deal with the threat of CB warfare. This attitude may have explained the many attempts to shut down Fort McClellan since the Gulf War.

In January 1991, the Army leadership recommended shutting down Fort McClellan and moving the Chemical School, claiming that the Army had no special need for toxic agent training such as at the CDTF. Based on the experiences of Gulf War soldiers, the Chemical

School convinced the Base Realignment Committee (BRAC) otherwise. In 1993, the Army proposed moving the Chemical School to Fort Leonard-Wood, Missouri, and keeping the CDTF open; it would fly trainees to Alabama for the specialized training. Again, the Chemical School successfully argued that the CDTF and the School should not be separated, due to the need to integrate up-to-date doctrine, training and equipment into the CDTF. In the 1995 recommendations to the BRAC, the Pentagon recommended the closures of both Fort McClellan and Dugway Proving Ground. The Army's argument to move the Schools (both Military Police and Chemical) was based on a new plan to develop an Army Maneuver Support Command, tieing in the Engineer Center and School with the two other schools to develop an integrated combat support concept. This time, the Army agreed to build a new CDTF at Leonard-Wood in addition to the move. The 1995 BRAC, on a very close (5-4) vote (with two Missouri politicians tipping the results), agreed to this plan.

Chemical officers remain concerned over the potential loss of the Chemical Defense Training Facility. The CDTF will remain active at Fort McClellan until one is constructed at Fort Leonard Wood-at least, as plans now stand. The CDTF has a very successful record of preparing soldiers, both chemical specialists and non-chemical personnel, for the all too real threat of a toxic environment. Tens of thousands of American soldiers have passed through the training, as well as hundreds of British, German, Canadian and other foreign military personnel. This was accomplished while maintaining a perfect record of no training days lost to maintenance or chemical agent accidents. Major General McCaffrey stated: "Then and now, they [soldiers and leaders] wholeheartedly value the opportunity to train with actual agents, real detectors/alarms and real decontaminants. The presence of CDTF-trained soldiers in every company of the Division directly improves our combat readiness. These soldiers have great confidence that their equipment works. Your training program is right on target."

While moving the Chemical School to Fort Leonard-Wood will not stop the Army's NBC defense program, it will disrupt the infrastructure and momentum of CB defense doctrine and training. Also, this plan assumes that there will be no delays in building a new CDTF in Missouri. The last CDTF was a five-year undertaking before the state of Alabama and the EPA approved its operations. Can we assume that Missouri will be more cooperative? If Fort Leonard-Wood cannot build a comparable CDTF by the time of the Chemical School will move at the turn of the century, there will be measurable drop in preparedness.

The attempted closure of Dugway Proving Ground is an even more serious issue. The loss of Dugway Proving Ground would eliminate the one place in the United States where extensive CB agent testing can still be accomplished. Without accurate testing of detection, protection and decontamination equipment with actual CB agents, how will the military be assured of good defensive equipment? As armored vehicles undergo live-fire tests, so must NBC defensive equipment undergo toxic agent tests. The argument for closing Dugway Proving Ground may have centered on the fact that since Congress drastically restricts open-air agent tests, the over 850,00 acres of desert are too expensive to maintain. Since testing with CB agents is all conducted indoors, the argument could be made to relocate the valuable test facilities elsewhere, although the cost would be high. The 1995 BRAC did feel that Dugway Proving Ground's unique

Testing facilities and proving grounds was vital enough to spare, and did not approve its inclusion on the closure list. Some type of ability to perform this testing is necessary. The army has not yet presented an effective argument as to who would assume the responsibility for this testing, and it may not have calculated the total cost of moving the test laboratories to other sites and shutting down the post.

I will not suggest any "*plots*" or "*conspiracies*" to disband the Chemical Corps (although three does seem to be a historical pattern). A cynic would point out that as long as the Infantry Ranger School trained its studies at Dugway Proving Ground, there were few recommendations to close the base (the Rangers left Dugway in 1992). Rather, I will point out that these actions take place without careful thought as to potential consequences. Yes, the Chemical School can still operate at another post. Yes, the testing facilities at Dugway Proving Ground could be replicated elsewhere. But no one has put any thought into minimizing the impact of these moves. Who will maintain the reforms in doctrine and training? Who will continue developing the NBC defense equipment requirements? Who will pick up the mission of testing military equipment to measure its ability to meet the errors of an NBC-contaminated environment? How much time and money will it cost the military to regain these capabilities? As seen with the DoD's attempts to model the Khamisiyah explosion at Dugway, there is still a role for this expertise. It Comes back to the need to maintain that high level of defense in a post-Cold War era, in a political environment that denies to the United States armed forces the use of retaliatory chemical weapons.

The Engineer Corps at Fort Leonard-Wood has a vision of a future combat support relationship. If there is one silver lining in the Chemical Corps move to the future Maneuver Support Command, it is that the Chemical Corps might emulate the successful "*teaming*" that engineers have developed with he combat arms branches. The cooperation seen between the Chemical and Engineer Corps in constructing deliberate decontamination sites in Southwest Asia suggests that there is room for both to grow together within that command. There is little question that the Chemical Corps could learn something from the Engineer Corps' ability to work closely with, and in support of, combat branches. There are two concerns foremost in the minds of chemical officers. History shows the many times that the US Army has attempted to nest the Chemical Corps within another branch (or submerging the potential role of a strong and vital Chemical Corps. The Chemical Corps's mission as a joint service leader, mandated by Congress, as well as its status of an international leader in NBC defense, mandates that it retain its autonomy to a degree. The US military cannot afford the Engineer Corps to assume to a degree. The US military cannot afford the Engineer Corps to assume the Chemical Corps mission within its own. What the Chemical Corps needs is a strong partner, not new management.

The Persian Gulf war should convince most that there is a strong need in this area for a force of full-time professionals. The military was clearly unprepared for CB warfare in 1990, and remains uncomfortable with the concept of CB warfare in 1990, and remains uncomfortable with the concept of CB warfare even today. The Army needs to study the future trends of warfare, analyze potential avenues for future investment and determine how to prepare for conflicts that will include NBC weapons employment. More importantly the Army needs the chemical Corps to develop doctrine and to recommend chemical organizations that fits the future Force XXI concept, and to develop leadership skills to ensure the entire military force has the capability to survive and sustain combat operations in an NBC-contaminated environment. To do this, the Chemical Corps must become a true partner with its sister branches of the Army, as well s its counterparts in the other services. In the past, the *Chemical Corps* has retained a technical specialty; for the future, it must become an integral part of the combat arms teams. The *Chemical Corps* must understand the combat arms' doctrine and leadership demands, and it must communicate its potential to the combat arms into terms their leaders understand in order to achieve these goals.

***In closing:***

Future US administrations must decide how the Department of Defense should respond to enemy use of CB munitions. The CWC exists, but treaties have never stopped nations from researching and developing more advanced CB agent munitions. Treaties have rarely stopped those same countries from employing CB agent munitions. Faced with the fact that the US military still has problems facing World War I and World War II-vintage chemical munitions, other countries do not have far to go to produce an effective weapons that will disrupt military operations. Arms control experts and politicians who rail against CB warfare research are living in the past. Their transition of the present must come in the realization that while they can strive to limit future warfare, they will never totally eliminate it. Countries find CB agent munitions just too tempting, too easy, to develop, too easy to hide, and too easy to deny having used, to pass up. When the military experts, politicians and arms control experts come to consensus on how best to integrate the trend of chemical warfare into future warfare policy, we can be assured that the United States has a sound national policy and a strong defensive capability in this area. Until then, our forces will remain dangerously vulnerable to CB weapons.

The underlaying reason why politicians, the military, and the public have feared and continue to fear, CB warfare so much is sheer ignorance about its nature. Either they completely underestimate its effects by ignoring the possibility of CB warfare during plans and exercises, or they completely overestimate the CB agents killing ability on the battlefield and in the eight stockpile. *Chemical* and *biological weapons* have never been on the same scale of destruction as nuclear weapons, yet many seem determined to treat them all the same way. At the least, chemical weapons have as much, and as little, potential killing power as any number of conventional bombs, artillery shells, and rockets in use today. Any weapon system is as safe or as deadly as the military force that uses them. As long as the media remains bent on sensationalistic journalism, and as long as novelists such as Tom Clancy and Hollywood screen writers make out CB weapons to be the doomsday devices *du jour*, we may never learn about the true threat of CB warfare to our own forces. Without understanding the tactical nature of these weapon systems, the military and government politicians remain unprepared to plan defensive strategies and will make the wrong decisions. This ignorance will continue until military leaders and politicians make an effort truly to understand the implications of CB warfare in modern combat operations.

We can no longer afford for this divisive battle of arms control experts versus the *Chemical Corps* versus the *combat arms* community. If there is a future vision for the *Chemical Corps*, it is ensure that combat forces can maintain their full mission capability in future wars despite the threat of a CB agent-contaminated battlefield. This call for both highly trained and motivated chemical officers trained and educated military leadership. As General Creighton Abrams once told Colonel Gerald Watson in 1972, it is the infantry and armor combat leaders—not its technical branches—that must decide how to fight and survive on the modern battlefield. The military may not like training for CB warfare, but the Chemical Corp's capabilities and CB defense equipment remain the sole "*insurance policy*" that maintains the United States military's ability to defend itself against the use of CB munitions. Until the *Chemical Corps* specialists are seen as true supporters of the combat soldier, and until the combat arms community comes to grips with the true scope and nature of CB warfare, the potential for future Desert Shield/Storm panics remains.

# INDEX


**A**

A.W. Bauer, 150
Acid-fast, 63
Acquire resistance, 151
Acridine dyes, 135
Acute disillusionment, 28
Aeromonas, 90
Aeromonas hydrophila, 90
Agar diffusion method, 150
AIDS, 64
Air Gas Defense, 2
Alcohols, 130
Alexander Fleming, 141
Alkylating agent, 131
Alpha-hemolytic streptococci, 54
Amblyomma, 103
Amikacin, 145
Aminoglycosides, 144
Amoxicillin, 143
Amphotericin B, 149
Ampicillin, 143
An unwarranted luxury, 28
Anaphylactic reaction, 142
Andromeda strain, 236
Animals, 81
Anionic detergents, 134
Anthrax, 6, 8, 12, 13, 14, 15, 17, 32, 42, 48, 92, 111
Anthrax aerosol tests, 30
Anthrax bomb, 18, 19
Anthrax infection, 30
Anthrax spores, 21
Anti-animal, 37
Antibiotic, 33, 137, 140
Antiseptic, 123
Antitoxins, 55, 75
Aplastic anemia, 146
Arnold sterilizer, 115
Arsphenamine, 138
Arthritis, 100
Aseptic, 124
Atomic bomb, 22
Atomic explosives, 22
Autoclave, 113
Axial filaments, 96
Azithromycin, 147

**B**

B. anthracis, 93
B. burgdorferi, 102
B. canis, 87
B. hermsii, 102
B. mellitensis, 87

B. pertussis, 59
B. recurrentis, 102
B. suis, 87
B. turicatae, 102
Bacille Calmette Guerin, 65
Bacillus, 114, 135, 148
Bacillus anthracis, 92, 223
Bacillus cereus, 89
Bacillus globigii, 31
Bacillus prodigiosus, 2
Bacitracin, 148
Bacterial dust, 9
Bacterial dysentery, 82
Bactericidal agent, 123
Bacteriostatic agent, 123
Bacteroides, 148
Bacteroides fragilis, 149
Bang's disease, 89
Battle Lab, 243
Bergey's Manual of Systematic Bacteriology, 70
Beta-hemolytic streptococci, 54
Beta-lactam nucleus, 142
Beta-propiolactone, 132
Big Tom, 38
Biological weapons, 5, 15, 20, 22, 34, 216, 255
Biological weapons agents, 9
Biological weapons programs, 42
Bioterrorism, 40, 48, 216
Bioweapons, 41
Bipolar staining, 98
Bisphenols, 128
Black Death, 97
Bleach, 127
Blunderbuss method, 30
Boiling water, 112
Booms and busts, 28
Bordetella pertussis, 57, 59
Born classified, 201, 206
Born secret, 203
Borrelia, 102
Borrelia burgdorferi, 100
Borrelia recurrentis, 102
Botulinum toxin, 8, 42
Botulism, 73
Botulus, 75
Bright line, 202
Brill-Zinsser disease, 107
Bronchopneumonia, 66
Brucella abortus, 87
Brucellosis, 87
Brucellosis, 20, 32
Buboes, 98
Bubonic plague, 97

C

C. botulinum, 75
C. burnetii, 70
C. coli, 87
C. difficile, 76
C. diphtheriae, 57
C. perfringens, 78
C. upsaliensis, 87
Calcium hypochlorite, 127
Campylobacter, 86, 87
Campylobacter jejuni, 87
Campylobacteria jejuni, 87
Campylobacteriosis, 86
Candida, 149
Candida albicans, 135, 146, 149
Carbapenems, 149
Carbenicillin, 143
Carcinogens, 178
Carriers, 80
Casualty production, 31
Catalase, 133
Cationic, 134
Cavitation, 119
Cavitron, 120
Cell, 212

Cephalosporins, 144
Cephalosporium acremonium, 144
Chemical, 255
Chemical and Biological Weapons, 210
Chemical Corps, 30, 254, 255, 256
Chemical Defence, 16
Chemical Defence Research Establishment, 5
Chemical weapons, 239
Chemotherapeutic agents, 137
Child, 26
Childbed fever, 56
Chlamydia, 147
Chlamydia pneumoniae, 71
Chlamydia psittaci, 71
Chlamydia trachomatis, 71
Chlamydial pneumonia, 71
Chloramines, 127
Chloramphenicol, 145
Chlorhexidine, 129
Chlorine, 127
Chloroquine, 140
Cholera, 82
Chromobacter violaceum, 148
Chronicum, 100
Clarithromycin, 147
Clavulanic acid, 144
Clear water, 127, 128
Clindamycin, 148
Clostridial food poisoning, 77
Clostridial myonecrosis, 95
Clostridium, 76, 95
Clostridium botulinum, 73, 75
Clostridium difficile, 148
Clostridium perfringens, 77, 95
Clostridium tetani, 93
Clostridium welchii, 95
Cloud chamber, 15
Codes of ethics, 191
Cold agglutinin screening test, 67
Cold War, 22, 241
Combat arms, 256
Competitive inhibition, 139
Consequence management, 248
Conspiracies, 253
Consumption, 63
Contagious abortion, 88
Contagious diseases, 32
Contentious research, 164
Control, 198
Copper, 130
Copper sulfate, 130
Corynebacterium diphtheriae, 56
Corynephage, 57
Countering Agricultural Bioterrorism, 171
Coxiella burnetii, 70, 116
Cresols, 128
Crisis management, 248
Cryptosporidium, 108
Cutaneous anthrax, 33

**D**

D.D. Woods, 139
Dangerous, 211
Dangerous substance, 195
Dapsone, 140
Deemed export, 211
Defense, 38
Degerm, 124
Denaturation, 112
Dermacentor, 103
Designer diseases, 164
Detergents, 134
Development, 211
Diatomaceous earth, 117
Diatoms, 117
Diphtheria, 56
Dipicolinic acid, 109
Direct flame, 111

Disinfectant, 123
Disinfection, 123
Domestic Preparedness, 248
Double pneumonia, 66
Drawbacks, 126
Droplets, 55
Dry heat, 111
Drying, 120
Du jour, 255
Dual use, 153, 198, 208, 215
Dysentery, 82
Dystonias, 76

E

E. coli, 85, 86, 144
E. equi, 107
E.M. Fildes, 139
Early warning, 228
Ectromelia, 164
Eggs, 81
Ehrlichia chaffeensis, 107
Ehrlichiosis, 107
Endemic typhus, 105
Enforcement, 180
Ensure, 220
Entamoeba histolytica, 145
Enteritis, 81
Enterotoxin, 76
Epidemic typhus, 105
Ernst Boris Chain, 142
Erysipelas, 56
Erythema, 100
Erythrogenic toxin, 55
Erythromycin, 147
Escherichia coli, 85, 152
Ethambutol, 140
Ethyl alcohol, 131
Ethylene oxide, 132
Exotic, 200
Exotoxin, 57, 73
Experiments of Concern, 227
Experiments of concern, 235
Exxon Valdez, 250

F

F. tularensis, 99
Filter, 116
First-generation, 144
Flaccid paralysis, 74
Flash pasteurization, 116
Flucytosine, 149
Fluoroquinolones, 140
Fodder disease, 75
Folic acid, 139
Formaldehyde, 131
Formalin, 131
Fractional sterilization, 115
Francisella tularensis, 32, 99

G

Gamma rays, 118
Gamma-hemolytic streptococci, 54
Gas gangrene, 8, 95
Gastroenteritis, 81
General purpose clause, 157
Gentamicin, 145
Gentian violet, 135
Gerhard Domagk, 138
Germ theory of disease, 162
Glomerulonephritis, 56
Glutaraldehyde, 132
Gold standard, 223
Gray syndrome, 146
Griseofulvin, 149

H

H. influenzae, 61
H. pylori, 86
Haemophilus, 61, 148
Haemophilus influenzae, 61, 144
Haemophilus meningitis, 61
Halogens, 126
Heat, 109

Heavy metals, 130
Helicobacter pylori, 86
Helleborus, 172
Hemolysin, 95
Hexachlorophene, 128
Hexylresorcinol, 129
Hibiclens, 129
High-efficiency particulate air (HEPA) filter, 117
High-risk, 192
His-Werner disease, 107
History, 105
Holding method, 115
Hospitals, 152
Hot oil, 116
Hot-air oven, 111
Howard Florey, 142
Human granulocytic ehrlichiosis, 107
Human monocytic ehrlichiosis, 107
Hyaluronidase, 95
Hydrogen peroxide, 133

I

I. dammini, 101
I. pacificus, 101
Ignaz Semmelweis, 121
Illustrative but not exhaustive, 210
Imidazoles, 149
Imipenem, 149
In absentia, 138
In vivo, 164
In-use test, 126
Inadvertent escalation, 1
Incapacitating, 32
Incubation period, 77
Indoctrination, 21
Infant botulism, 76
Infantile diarrhea, 85
Infections, 73
Infectious agents, 178
Inorganic filters, 117
Insurance policy, 256
Interim, 34
Intermittent sterilization, 115
Intoxications, 73
Intravenous injections, 83
Iodine, 127
Iodophors, 128
Ionizing radiations, 118
Isoniazid, 140
Isopropyl alcohol, 131
Ixodes, 100
Ixodes scapularis, 101

J

Jacques and Therese Trefouel, 138
Joseph Lister, 123
Journal of the Plague Year, 97
Journal of Virology, 165, 217
Julius Petri, 116

K

Kanamycin, 145
Kirby-Bauer test, 150
Klebsiella, 68
Klebsiella pneumonia, 67
Klebsiella pneumoniae, 67

L

L. pneumophila, 69
La diphtherite, 56
La tache noire, 107
Large Area Concept, 36
Laser beam, 118
Law of Land Warfare, 35
Lecithinase, 95
Legionella, 69
Legionella pneumophila, 53, 69
Legionellosis, 69
Legionnaires' disease, 69
Leptospira interrogans, 95
Leptospirosis, 95
Lethal, 28

Lethal chemical, 22
Lice, 105
Limberneck, 75
Limitations, 114
Lincomycin, 148
Listeria monocytogenes, 96
Listeric meningitis, 96
Listeriosis, 96
Livestock feeds, 152
Lobar pneumonia, 66
Lockjaw, 94
Looking-glass, 3, 17
Low temperatures, 120
Luft-Gas-verteidigung, 2
Lyme disease, 100

**M**

M protein, 55
M. pneumoniae, 67
M. tuberculosis, 64, 133
Macrolides, 147
Maculopapular rash, 103
Magic Sword, 38
MAI disease, 65
Major actions, 181, 183
Malta fever, 89
Manage, 198
Materials and methods, 165
Melioidosis, 97
Membrane filter, 117
Meningitis, 9, 59
Meningococcal meningitis, 59
Meningococcemia, 60
Meningococcus, 59
Mercury, 130
Metachromatic granules, 56
Metronidazole, 140
Mexican typhus, 106
Micromonospora, 145
Microwave, 118
Migrans, 100
Miliary tuberculosis, 63
Milium, 63
Minimum inhibitory concentration, 150
Misuse of biotechnology, 236
Misused, 218
Mode of transmission, 77
Moist heat, 112
Monobactams, 148
Montezurna's rcvenge, 85
Mousepox, 211
Moxalactam, 148
Multi-drug-resistant Mycobacterium tuberculosis, 64
Murine, 106
Murine typhus, 106
Mushi, 106
Mycobacteriosis, 65
Mycobacterium, 65, 140
Mycobacterium avium-intracellulare, 65
Mycobacterium bovis, 64
Mycobacterium cheloni, 65
Mycobacterium haemophilum, 65
Mycobacterium scrofulaceum, 65
Mycobacterium tuberculosis, 62, 63, 116
Mycoplasma pneumonia, 66, 67

**N**

N. meningitidis, 60
Naive, 158
Nalidixic acid, 140
Nature, 212
Necrotizing fasciitis, 56
Need to know, 206
Neisseria, 61, 147, 148
Neisseria gonorrhoeae, 130, 143
Neisseria meningitidis, 59, 60
Neomycin, 145
New World Order, 241

Nitrofurantoin, 140
No country, 197
No first use, 238
Nonlethal diseases, 20
Nosocomial, 68
Novel, 215, 236
Nystatin, 149

**O**

Official Use Only, 210
Operation Aladdin, 12
Operation Vegetarian, 12
Organic acids, 135
Organic filters, 117
Ornithodorus, 102
Ornithosis, 71
Osmosis, 120
Otitis media, 55
Ownership, 192

**P**

P. shigelloides, 90
P. tularensis, 37
Para-aminobenzoic acid, 139
Para-aminosalicylic acid, 140
Parameters of chemistry, 125
Parent, 26
Paromomycin, 145
Parrot fever, 8
Pasteurella pestis, 98
Pasteurella tularensis, 33
Pasteurization, 115
Paul Ehrlich, 137
Paul Vuillemin, 140
Pediculus, 105
Penicillin, 141
Penicillin G, 142
Penicillinase, 142
Penicillium, 141, 149
Pertussis, 57
Phenol, 128
Phenol coefficient, 126
pHiso-Hex, 128
Plague, 12, 32, 42
Pleomorphic, 67
Plesiomonas shigelloides, 90
Plots, 253
Pneumococcal pneumonia, 65
Pneumococcus, 65
Pneumoncystis, 139
Pneumonia, 65
Pneumonic plague, 98
Polymyxin B, 148
Pontiac fever, 69
Port shopping, 43
Potential users, 15
Pott's disease, 63
Poultry products, 81
Practicability, 12
Prevacuum autoclave, 115
Primaquine, 140
Primary atypical pneumonia, 66
Primary disease, 55
Proceedings of the National Academy of Sciences, 168, 212
Production, 211
Proliferators, 194
Prontosil, 138
Proteus, 104, 152
Proteus OX19, 103
Provocative, 163
Pseudomembrane, 57
Pseudomembranous colitis, 148
Pseudomonas, 145
Pseudomonas aeruginosa, 135, 140, 144, 148, 152
Pseudomonas cepacia, 128
Pseudomonas pseudomallei, 97
Psittacosis, 71
Psittakos, 71
Ptomaines, 76

Public revulsion, 18
Puerperal sepsis, 56

**Q**

Q fever, 32, 38, 70
Quaternary ammonium compounds, 134
Quinine, 140

**R**

Rabbit fever, 33
Rabies, 8
Radioactive material, 22
Rats, 105
Red Admiral, 25
Red Cloud, 38
Rehydration, 82
Relapsing fever, 102
Resistance, 78
Respiratory challenges, 33
Restricted person, 176, 186
Rheumatic fever, 55
Rheumatic heart disease, 56
Rice bowls, 243
Rice-water stools, 83
Rickettsia akari, 106
Rickettsia conorii, 107
Rickettsia prowazekii, 105, 107
Rickettsia rickettsii, 103
Rickettsia tsutsugamushi, 106
Rickettsia typhi, 106
Rickettsiae, 70
Rickettsialpox, 106
Rifampin, 148
Roadmap, 234
Rochalimaea quintana, 107
Rogue, 213
Rogue states, 153
Rose spots, 78

**S**

S. aureus, 77, 133
S. enteritidis, 80
S. heidelberg, 80
S. newport, 80
S. pneumoniae, 66
S. poona, 73
S. pyogenes, 55, 56
S. typhi, 78, 79, 80
S. typhimurium, 80
Sahachiro Hata, 138
Salmonella, 72, 73, 76, 79, 80, 81, 87, 236
Salmonella poona, 72
Salmonella typhi, 78, 126
Salmonellosis, 80
Salting, 120
Salvarsan, 138
Sanitize, 124
Scarlet fever, 55
Science, 68, 69, 166, 212
Scientific Communication and National Security, 171
Scrofula, 65
Scrub typhus, 106
Second-generation, 144
Secrecy orders, 206
Select agents, 227, 231
Selman A. Waksman, 145
Semipersistent, 247
Sensitive, 166, 199, 204, 208, 209, 212, 214, 235
Sensitive but unclassified, 208, 216, 226
Sensitive but unclassified information, 172, 209
Sensitive homeland security information, 210
Sensitive information, 208
Sensitive Subjects List, 210
Sepsis, 124
Septicemia, 56
Septicemic plague, 98
Serratia marcescens, 2, 31, 68, 152

Serratia pneumonia, 68
Seventh Day Adventist, 33
Severe acute respiratory syndrome, 222
Sh. boydii, 82
Sh. dysenteriae, 82
Sh. flexneri, 82
Sh. sonnei, 82
Shady Grove, 38
Shigella, 81, 82, 86
Shigella dysenteriae, 152, 183
Shigellosis, 82
Shipboard Hazard, 38
Silver, 130
Silver nitrate, 130
Smallpox, 32, 42
Soap, 133
Sodium hypochlorite, 127
Spanish Flu, 222
Spectinomycin, 148
Spores, 109
Sputum, 62
Staccato coughs, 59
Staphylococcal enterotoxin, 39
staphylococcal food poisoning, 76
Staphylococcus, 135
Staphylococcus aureus, 39, 76, 77, 126
Sterilization, 108, 123
Sterilizing chamber, 114
Strep throat, 55
Streptococcus, 121
Streptococcus pneumoniae, 65, 66, 67
Streptococcus pyogenes, 55
Streptomyces, 145, 146, 147, 148, 149
Streptomyces griseus, 145
Streptomyces venezuelae, 146
Streptomycin, 145
Sulfamethoxazole, 139
Sulfanilamide, 139
Sulfisoxazole, 139
Sulfonamides, 139
Surveillance, 47
Swap meet, 230
Sylvatic plague, 98

**T**

Tabardillo, 106
Tabardo, 106
Teaming, 254
Technology, 190, 201
Tetanospasmin, 93
Tetanus, 8, 93
Tetracyclines, 146
The Bible, 122
The Citadel, 109
The Lancet, 142
Thermal death point, 110
Thermal death time, 110
Third World countries, 152
Third-generation, 144
Thymine molecules, 117
Ticarcillin, 144
Tickborne fevers, 107
Tincture of iodine, 127
Tobramycin, 145
Top officials, 49
Toxoid, 57, 94
Traveler's diarrhea, 85
Trench fever, 107
Treponema pallidum, 138
Trimethoprim, 139
Triphenylmethane dyes, 135
Trombicula, 106
Tsutsuga, 106
Tsutsugamushi, 106
Tube dilution method, 150
Tubercle, 63
Tubercle bacilli, 110
Tuberculin test, 64

Tuberculosis, 61
Tularemia, 6, 32, 33, 38, 39, 42, 99
Tyndallization, 115
Typhoid, 9
Typhoid fever, 78, 109
Typhos, 105
Typhus fever, 105

U

U.S. Pharmacopoeia, 122
Ultrapasteurization, 116
Ultrasonic vibrations, 118
Ultraviolet light, 117
Unclassified controlled nuclear information, 201
Undulant fever, 88
Unintended consequences, 236
Unsoldierly, 17

V

V. cholerae, 83, 84
Vaccinia, 164, 165, 168
Value-added, 244
Vancomycin, 147
Variola, 168, 184
Variola major, 168, 184
Vector, 98
Vehere, 98
Venezuelan equine encephalitis, 33
Vibrio cholerae, 83
Vibrio purahaemolyticus, 89
Vigilance Committee, 9
Viking Serenade, 82
Visible light, 117

W

W.M. Kirby, 150
Walking pneumonia, 67
Waterhouse-Friderichsen syndrome, 60
Weapons of mass destruction, 243, 244
Wears the hat, 242
Weil-Felix test, 103
Wetting agents, 133
Whooping cough, 9, 57
Widal test, 78
Woolsorter's disease, 93
World War I, 250
World War II, 239, 250
Wound botulism, 75

X

X rays, 118
X. fastidiosa, 156
Xenopsylla cheopis, 98, 105

Y

Y. enterocolitica, 90
Y. pestis, 98, 99
Yellow Leaf, 38
Yersinia enterocolitica, 89
Yersinia pestis, 98, 158

Z

Zoonosis, 95